Understanding
the
Astrological Vertex

Understanding the Astrological Vertex

Love & Life when Fate takes the Driving Seat...

Sasha Fenton

Zambezi Publishing Ltd

Published in 2006 by
Zambezi Publishing Ltd
P.O. Box 221 Plymouth,
Devon PL2 2YJ (UK)
web: www.zampub.com email: info@zampub.com

Copyright: © 2006 Sasha Fenton
Cover design: © 2006 Jan Budkowski

Sasha Fenton has asserted the moral right
to be identified as the author of this work in terms of
the Copyright, Designs and Patents Act 1988

British Library Cataloguing-in-Publication Data:
A catalogue record for this book is available from
the British Library

Typeset by Zambezi Publishing Ltd, Plymouth UK

(ISBN-10):   1-903065-12-7
(ISBN-13):   978-1-903065-12-9

135798642

About the Author

A career as a professional astrologer, palmist and tarot card reader began in 1974. Sasha's first book (Fortune Telling by Tarot Cards) was published in 1985, and she has now written an awesome figure of more than one hundred and twenty titles, mainly on mind, body and spirit subjects. With her easy-to-read writing style and total sales now exceeding seven million copies, Sasha is clearly an outstanding authority on a range of fascinating subjects.

Sasha's contributions to the fields of writing and mind, body and spirit include past service as a member of the Executive Council of the Writers' Guild of Great Britain, Chair of the Advisory Panel on Astrological Education, and currently as President of the British Astrological and Psychic Society (BAPS) for the second time.

At present, Sasha devotes most of her time to running Zambezi Publishing Ltd with her husband, Jan Budkowski., through which they help in presenting new and deserving talent in the world of spirit.

Acknowledgments

To Jonathan Dee, who rooted around in his huge library and dug out the method of calculating the co-latitude for me.

To Sean Lovatt, for technical advice on many astrological features over the years.

Thanks to Jan for the fiddly job of fitting the charts and diagrams into this book, and for helping me with some of the research.

Thanks to all the unwitting pop stars, actors, politicians and the occasional mass-murderer whose charts help bring my books to life.

Contents

1

Introducing the Vertex

The Vertex is one of the most fascinating topics in astrology, but if you are to interpret it for yourself, you will need a copy of your birth chart. At the very least, you will need to know the sign and house that your Vertex occupies. Once you have the chart, look for the symbol **Vx** on it, as that will be your Vertex.

There are various ways of going about getting a chart. One starting point is my website, *www.sashafenton.com*, which will show you how to access various free or inexpensive chart services. Other than that, you can consult any astrologer who uses a computer and professional astrological software, or who can follow the hand calculations that are described in the relevant chapter in this book. As it happens, the calculation is extremely easy, so any astrologer who uses normal astrological systems will be able to work it out quickly for you. If you are a trainee astrologer, you will be able to work out the Vertex position yourself without much difficulty, and the Anti-Vertex is always opposite the Vertex, so this is easy to spot.

In the next chapter, you will find a sample chart that shows the Vertex and various other points that I discuss in this book. After that, you will see lots of charts, so by the time you have finished this book, you will be at home with the technicalities – none of which are particularly difficult to understand.

One idea that you might wish to pursue is to buy a copy of this book and give it to an astrologer of your acquaintance. Allow him time to absorb the technique, then ask him to explain some of the more complex points to you, such as the aspects to the Vertex and predictive techniques. Believe me, it won't take your astrologer any time at all to get the hang of the Vertex and Anti-Vertex and to use the knowledge for your advantage – and his.

Just for a change, we have written this book addressing a feminine readership. No significant point intended, just a change!

2

About the Vertex

No serious astrologer can possibly believe in fate! None of us can live with the idea that some spiritual puppet master is pulling our strings. We govern our own destinies, don't we?

Then there is the Vertex...
The Vertex is a sensitive point on the birth chart that concerns attraction, desire, obsession, falling in love, powerful family relationships and even important working associations and friendships. It can show how or why we form friendships or what it is about others that fascinates us. It can show why our subconscious makes the choices that it does. It shows what attracts us to others, what attracts them to us and why our emotions sometimes have a mind of their own. The Vertex is activated when something important happens and when it upsets our equilibrium.

If things have been going along in a routine or an unsatisfactory manner for some years, our unconscious minds gradually prepare us for change. Perhaps our spiritual guide or the "universe" takes a hand - then an eclipse or some other astrological event hits the Vertex or Anti-Vertex, and suddenly we take off to pastures new. In its heaviest forms, the Vertex can feel similar to Pluto in its effect, but it can also mimic Venus, the Moon, Mars, Jupiter and Uranus. Indeed, it can shadow any planet that is able to put our beliefs and feelings through the wringer.

Minor transits to or by the Vertex signal minor events, such as pleasant short-lived friendships, pleasant flirtations or a momentary feeling of understanding between a perfect stranger and ourselves. These can also account for unexpected arguments with strangers – such as coming across an intransigent traffic warden or someone who pushes in or cuts us up on the road.

For many of us, finding a partner is a major preoccupation. We want to know where, when and how we will meet our soul mate, so studying the Vertex might help us to discover the best time to be open to new influences. On the other hand, we find it unbearable to split from someone who means the world to us; so watching events around the Vertex can warn of impending storms and possibly help us to avoid them. Relationships can make us happy or miserable, and it is mainly problems linked to love that drive us to consult psychics, astrologers and Tarot readers. Love can make our lives worthwhile, or it can quite literally break our hearts.

Beyond our intimate adult relationships, we interact with family members, friends, bosses, colleagues and neighbours on a daily basis. Even here, we can feel fulfilled by great relationships, angered and hurt by bad ones or totally devastated by the death of a much loved relative or friend. The Vertex will be involved in many cases of this kind.

True Stories About Love

Joanne escaped a dysfunctional family home by marrying while she was still a teenager. Needless to say, the marriage was a mess. While bringing up her children, she had a long affair with a man who refused to leave his wife. Eventually Joanne found the strength to leave this man, as well as leaving her violent husband. She then met another married man who also wouldn't leave his wife. Eventually, Joanne dumped this man and found someone who was actually available to her and she married for a second time. The binding factor in this second marriage is the couple's mutual love of alcohol. Joanne's Vertex is in Virgo in the eighth house. Her Anti-Vertex is in Pisces in the second house. Her first husband was a Pisces, her second was a Virgo and her two unsatisfactory lovers were both Piscean.

*

Derek's mother, sister, wife and children were Leos, as were his two best friends. His Vertex was in Scorpio. He found a job with a firm called Taurus Construction, but couldn't get on with the boss or their slow and pedantic way of doing things. His Anti-Vertex was in Taurus!

*

If you have lived through an intense relationship that has been hugely successful, or one that has exerted a terrible emotional (and financial) cost, the Vertex and Anti-Vertex are worth looking at. They may show how you set yourself up for pain, or they may show some form of synastry between you and your lover. If you keep an eye on your Vertex, you can even plot those times when you are most likely to fall in love or be vulnerable to emotional attack.

*

Some years ago, a client of mine fell in love with a man over the phone. She was driven insane by lust for this man, and when she finally met him and started an affair, her feelings deepened to the point at which they totally overwhelmed her. The strange thing was that my client knew that the man was no good for her, and probably not much use to any other woman either. Even though she knew this, she couldn't stop her heart and her emotions taking over, and the resultant turmoil eventually broke up her marriage. The man was only really after my client's money, and he helped himself to a fair amount of it before the sorry business came to an end.

My client's first encounter with this man happened when a total solar eclipse fell on her Vertex. In reality, the affair ended a few months later, when a partial lunar eclipse occurred near her Anti-Vertex, but my client hung on, eating her heart out and continuing to pursue the detestable man. She finally came to her senses when another partial solar eclipse occurred in the same sign as her Vertex (although not right on it). After this, she gave herself time and space to discover who she really was, and what she wanted out of life, before looking around for the next relationship.

*

Another client of mine had been happily single for some years, but then she met a nice man who surprised her by asking her to marry him. On that day, there was a solar eclipse on his Vertex!

Factors Other than Romantic Ones

Watch the Vertex at times of bereavement. See if it progresses to a sensitive point or if some other factor on the chart makes progressions or transits to it. Serious illness is another Vertex issue, as it brings worry and extra responsibilities to various other people who surround the sick person.

*

Some astrologers have discovered that the Vertex connects with work: not so much the kind of work that we do, but the results that arise after we have made an effort. This is probably most common in those whose Vertex is in the sixth astrological house, which focuses on duty and employment. Business connections to people of one's Vertex sign seem to work well for most of us. Sometimes oppositions and eclipses to the Vertex throw one's career out of the window, and they can even result in someone being unjustly accused of something that they didn't do.

*

A person's belief system can prove to be faulty. A guru can be shown to be empty or to have questionable motives. On the other hand, a person can suddenly discover something worth believing in, or he might come across a good teacher and guide. Either way, the condition of the Vertex at the time of change is worth considering.

The Vertex, Anti-Vertex, East & West Points

Does the Vertex Exist?

There are astrologers who say that the Vertex doesn't exist, but they are wrong; it definitely does. The prime vertical cuts through the ecliptic, that's the Vertex, and that's that.

Does Fate Exist?

Some scientifically minded astrologers dislike the idea of fate, but they probably haven't yet experienced life-changing events, severe illness, falling in love or meaningful bereavement. When they do, they will understand the life-changing effects of destiny.

The East and West Points

Close to the Ascendant and Anti-Vertex is the East Point. This is the point where the eastern horizon crosses the celestial equator. It is similar to the Ascendant, which itself is the place where the eastern horizon meets the ecliptic. Some astrologers have called the Anti-Vertex and the East Point hidden or secret Ascendants.

Before you wonder why the East Point and the Ascendant should be on the left hand side of the chart, it is worth considering the heavens in the way that the ancients did, and using their logic when looking at an astrological chart.

Take your chart outside around midday and face towards the south. The MC will point towards the Sun and the Ascendant and East Point will point to the east, which is where the Sun rises. Obviously, the West Point and Descendant will be pointing west towards the direction where the Sun sets.

If you happen to live in the southern hemisphere, then you must point your MC north towards the Sun and turn your chart face down. That way, the Ascendant will be in the east, along with the East Point.

A Sample Chart

If you have software, select the Vertex and East Point so that they show up on your charts. The symbol for the Vertex is **Vx**, and the symbol for the East Point is sometimes **Ep** and at other times **Eq**. The reason for this is that some astrologers (and some software systems) refer to the East Point as the *Equatorial Ascendant*.

An Example

I have given you the following example to show you what to look for on your own chart. This example happens to be the chart for the glamourous Spanish actor, Antonio Banderas.

You will find Antonio's Vertex at 19 degrees Virgo in the seventh house on this chart. The Anti-Vertex is always directly opposite the Vertex, but it isn't marked on a chart, so you might like to draw it in for yourself. In the case of Antonio Banderas, the Anti-Vertex is at 19 degrees Pisces in his first house. Antonio's East Point is

13 degrees Pisces and his West Point is 13 deg. Virgo. There is a gap of only six degrees between Antonio's East Point and Anti-Vertex, and six degrees between his West Point and Vertex.

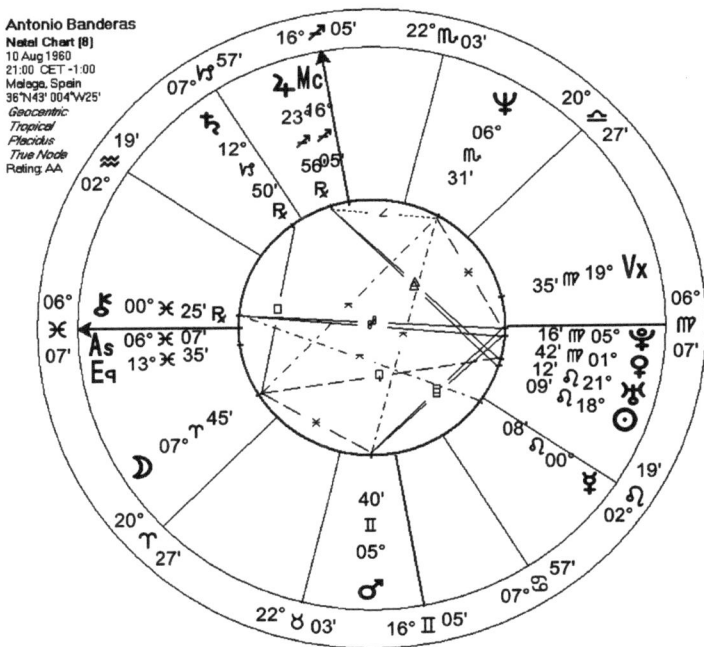

Note: Only the Vertex, East Point (Equatorial Ascendant) and Ascendant are usually shown on a chart; the Anti-Vertex, West Point and Descendant are directly opposite, and not marked, for reasons of clarity.

4
Technical Data

If you happen to be a mathematician, you will be overjoyed to know that the Vertex is simply the point where the prime vertical cuts though the ecliptic. The ecliptic is the apparent trajectory of the Sun around the earth. The Greenwich meridian is the line that passes, north to south, through Greenwich and around the earth. This line is called a great circle. The prime vertical is a similar great circle that runs, also vertically, at right angles to the Greenwich meridian. The Vertex is the place where this line crosses the ecliptic.

Like all astronomical features, the Vertex has been known for many years, but early in the 20th century, a man called Edward L Johndro decided to look into it. Johndro was a physicist, mathematician, radio engineer and astrologer. A century or so ago, scientists were fascinated by electricity, so they looked into all kinds of new explanations in the light of its discovery. Johndro was seeking a scientific explanation for astrology by looking at electrodynamics. Like most astrologers, Johndro didn't make the kind of breakthrough that would impress the sceptical world of science, but the longer he studied the Vertex, the more he became convinced that it was important in the area involving personal relationships.

Johndro considered the Anti-Vertex (the point opposite the Vertex on a chart) to be an "electric ascendant". I have also heard this called the hidden or secret ascendant. Further study of the Vertex and Anti-Vertex axis led him to believe that this drew people together in a fated manner. Astrologer Charles Jayne considered the Vertex to show relationships and other matters that are beyond our conscious control. My own investigations suggest that both ends of this axis (both the Vertex and the Anti-Vertex) are involved in this process.

The Vertex appears on the "western" or right hand side of the chart, and it is usually in the astrological houses numbered five, six, seven and eight. It is possible for it to appear in the fourth or ninth house, but this is very rare. If you have the kind of software that allows you to watch the planets in motion, set up any chart on any day and set the timer to run hour-by-hour fairly slowly. You will see the Vertex climb up the right hand side of the chart, turn and then climb down again, then turn again and climb again, basically swishing up and down in an arc on the right hand side of the chart.

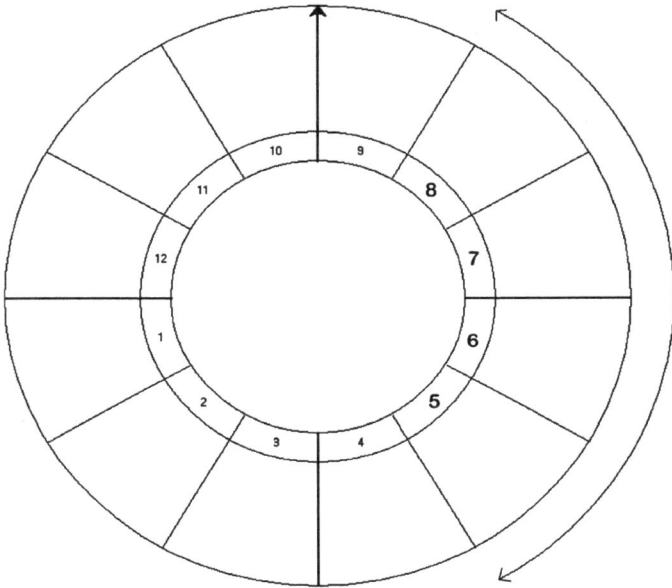

The Vertex normally traverses Houses 5,6,7 & 8

The opposite point to the Vertex is the Anti-Vertex. Just as the Vertex is always on the western side of the chart, the Anti-Vertex is on the eastern side of the chart. Software doesn't show the Anti-Vertex, but it is easy to work out, because it is always directly opposite the Vertex.

Now, for the fun of it, make up a chart for a tropical or polar region and run the clock. You might find the Vertex running up and down the wrong side of the chart! This is because the system gets confused and swaps the Anti-Vertex with the Vertex. Astrologers who use hand calculations get equally confused, and if you want to see just how confusing this can be, read Robert Hand's book, "Essays in Astrology".

Calculating the Vertex

There are very few astrologers these days who know how to calculate an Ascendant by hand, but for those who do, this is the procedure for the Vertex. I suggest that you only try this for someone born in the northern hemisphere in a temperate zone, as anything else is likely to be inaccurate.

Method

Deduct the latitude of birth from 90 degrees – this becomes the co-latitude. Take your Table of Houses and turn to the page headed by a latitude that is as close as possible to your co-latitude. Find the new MC (the cusp of the tenth house) in the table. Then, track across to the Asc. (the cusp of the first house), and the figure there will be the position of the Vertex.

You don't have to consider the longitude, so it doesn't matter whether the person is born in Europe, the USA, India or any other time zone.

Example One

Prince Charles:
14 November 1948. 21:14 London (51N30)

```
90 00
51 30 –
38 30
```

Nearest latitude:	Lisbon (38N43)
Charles's natal MC:	13 deg. 16 min. Aries.
Convert:	13 deg. 16 min. Libra.
New ascendant:	22 deg. 29 min. Sagittarius

Therefore, Charles' Vertex is 22 deg. 29 min. Sagittarius

Incidentally, when using a computer, the Vertex is 22 deg. 51 min. Sagittarius, which shows that the hand method is pretty accurate.

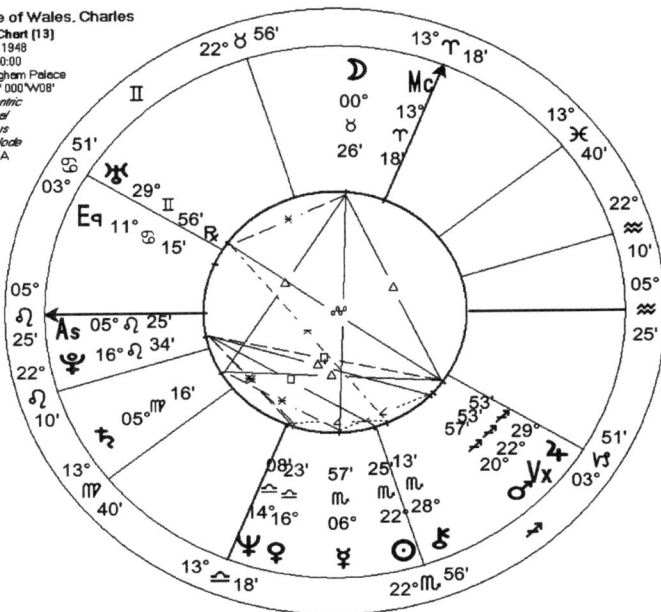

Prince of Wales. Charles
Natal Chart (13)
14 Nov 1948
21:14 +0:00
Buckingham Palace
51°N30' 000°W08'
Geocentric
Tropical
Placidus
True Node
Rating: A

Example Two

Bill Clinton: 19 August 1946. 08:51 CST (+6:00). Hope, Arkansas (33N40)

> 90 00
> 33 40 –
> 56 20

Nearest latitude:	Prague (50N50), Leningrad (59N56)
Bill's natal MC:	5 deg. 59 min. Cancer
Convert:	5 deg. 59 min. Capricorn. (Round this up to 6 deg. Capricorn)
New ascendant:	at Prague > 14 deg. 38 min. Aries.
	at Leningrad >26 deg. 58 min. Aries.

A rough assessment would make this 20 deg. Aries.

When worked out by computer, this comes to 19 deg. 52 min. Aries. As you can see, the hand method is remarkably accurate.

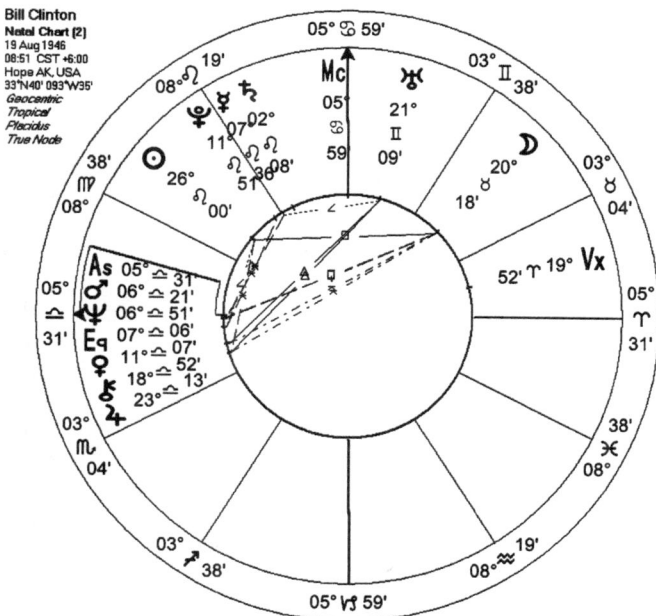

Bill Clinton
Natal Chart (2)
19 Aug 1946
08:51 CST +6:00
Hope AK, USA
33°N40' 093°W35'
Geocentric
Tropical
Placidus
True Node

5

The Vertex and the Houses

After much research, Jan and I discovered that the Vertex practically always occupies the fifth, sixth, seventh and eighth houses. We have found one chart that has it in the fourth house. Where discrepancies show up, they are due to the peculiar way that the maths works when one reaches the poles or the equator. If you want to know more about this, I suggest that you read Robert Hand's book "Essays in Astrology".

Just to show you how complex things can get in astrology, read the strange tale of Olga, the Russian artist. Jan and I met Olga several years ago while we were at a gathering of astrologers and psychics in London.

Olga's Story

"I 'ave no Ascendant," breathed the charmingly accented foreign voice in my ear.

"What?" I replied in surprise. "What do you mean, you have no Ascendant?"

"I am Russian and my name is Olga", said the attractive young lady; "I was born way up in the north of Russia during polar night, and though I 'ave been to many Russian astrologers and even one or two 'ere, nobody has been able to find my Ascendant."

This was certainly an unusual situation and it looked as though Olga might be right. In popular astrological systems, such as Placidus and Equal House, dropping a line down from sunrise at the birth latitude to the ecliptic forms the Ascendant. So, in theory, if there is no sunrise, there can be no Ascendant!

I was intrigued, so I said, "you'd better come and see me and I'll think of something."

NB: Olga was born on the 21st January 1963 at 11:10 a.m. local time in Norilsk, Northern Russia.

A couple of weeks later, Olga came for her consultation, and as luck would have it, my friend, Sean Lovatt, one of Britain's most technically knowledgeable astrologers, happened to phone me. I told Sean about Olga and he stayed on the line while entering her details into his computer. At the same time, I put her details into mine. Sean and I experimented with various different house systems. Sean suggested that we try Meridian or Campanus, both of which use the MC/IC line, and then he tried Campanus and commented that it showed an Ascendant of 27 degrees of Virgo. He also muttered darkly that in theory, Olga's Ascendant was also her descendant. Sean cut the connection and went off to play with Olga's chart by himself.

The phone rang again and this time my friend, Jonathan Dee, came on the line. Jon is another very experienced and talented astrologer, and he immediately threw Olga's data into his machine and said that he would ring back if he came up with any bright ideas. Before he rang off, Jon said that he would give

Porphyry a try. Meanwhile I got on with trying anything that I could think of.

I discovered that Porphyry showed four astrological houses, these being the first, sixth, seventh and twelfth. Campanus was better, as it came up with the same general picture as Porphyry, but it did show twelve shrivelled houses gathered together at the MC and the IC. Calls came in from Jonathan and Sean, both saying that they had also discovered that Campanus gave the best results. It was interesting that Sean, Jon and I were all using different software. While none of the programs could cope with this chart, at least they all made exactly the same mess of it!

Olga understood the system, but she commented that the Virgo Ascendant that we had all found simply didn't fit her personality. I was just coming to the crazy conclusion that Olga's chart was upside down when Jon phoned me back to tell me that he had also come to that conclusion! Then Sean phoned to tell me that the chart was not only upside down but also back to front! A Virgo rising chart showed the Sun and its close neighbours, Mercury and Venus, in the lower hemisphere, but Olga was born at 11.10 am. Even allowing for Polar night, a morning birth couldn't possibly put the Sun in the lower hemisphere: i.e. below the equator!

For people like us who live in the northern hemisphere, the only way to make sense of an astrology chart is to stand facing the south. Then the Midheaven points towards the Sun, the Ascendant is on the east and the descendant on the west. If I did this with Olga's chart, it showed that the Sun was shining above the North Pole – or more precisely, around the other side of the earth. Clearly, at that time of the year, it was – but this made no astrological sense.

Olga's computer generated chart showed aspects that weren't actually there. Stranger still, the nodes of the Moon were in trine to each other rather than in opposition! The chart bore some resemblance to older types of European charts in which the houses stayed equal and the signs stretched to fit them, but the planets, the nodes and the signs were all so far out of alignment that it was hard to see what the aspects really were.

I gave up on the astronomical impossibilities and followed a hunch that Pisces was the probable Ascendant. I gave Olga the information about the personality and the early life experiences of someone with Pisces rising. Olga agreed with my description of this Ascendant and told me that her childhood situation was exactly how I had described it. I then rectified the chart to give an Ascendant of 27 degrees Pisces and set it to the Equal House system. This at last gave me a workable chart and it was no hardship to pencil in the correct position for the Moon on both the natal and progressed charts. I then interpreted the chart for Olga in the normal way, taking my time about it and looking at every point in detail. An ecstatic Olga went away with a bright, shiny, new Ascendant.

The Present

Software has improved beyond all measure since those days, but some house systems still fall down in such circumstances. A Placidus chart works well enough. It keeps the Sun above the horizon and the Ascendant in Pisces, but it has somehow morphed itself into becoming Equal House! And look at the Vertex! It is next to the East Point... and that is impossible! However, both computerised calculations and hand calculations often confuse the Vertex and Anti-Vertex in these latitudes, so the thing to do here is to swap the Anti-Vertex for the Vertex.

Olga
Female Chart [2]
21 Jan 1963
11:10 -7:00
Norilsk, Russia
69°N20' 088°E06'
Geocentric
Tropical
Placidus
True Node
Rating: A
Mother

Placidus houses that look like equal houses. Note the Vertex shown, which is actually the Anti-Vertex. Note the MC at the <u>bottom</u> of the chart!

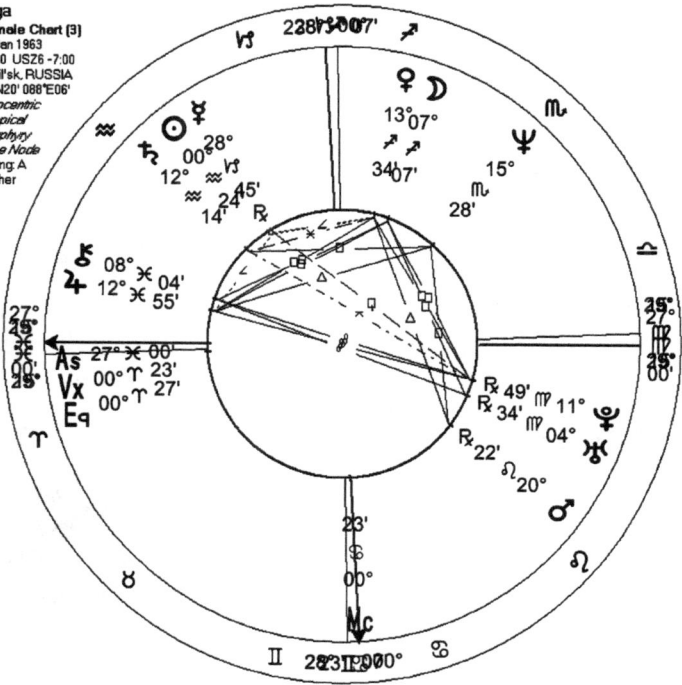

Olga
Female Chart (3)
21 Jan 1963
11:10 USZ6 -7:00
Noril'sk, RUSSIA
69°N20' 088°E06'
Geocentric
Tropical
Porphyry
True Node
Rating: A
Mother

This shows how Porphyry houses tried to cope. What is shown as the Vertex is, of course, the Anti-Vertex, and the MC is the IC!

Vx

6

The Vertex through the Signs

Vertex in Aries

If your Vertex is in Aries, your Anti-Vertex is in Libra. These are both masculine, cardinal signs, which means that you are unlikely to make too many compromises for the sake of your lovers, although you may learn to do so later in life after a few failed relationships. Aries is a fire sign while Libra is an air sign, so your natural inclination is to barge ahead with what you think is right, although you worry that you'll live to regret it.

The common thread here is "self" versus others, because Aries is concerned with personal freedom, self-motivation and self-centredness. Libra is concerned with companionship, sharing the workload and doing everything with someone else in tow. I remember a Libran once telling me that, if he decided to wash his socks, he would scrub them while his wife wrung them out for him or vice versa, because in his estimation, sock washing as a solitary occupation was just too lonely a thing to contemplate. The Arien would be horrified at such an idea. If absolutely necessary, he would wash his socks himself, but more likely, he would get someone else to do it while he went out for the day!

You fall in love in an instant and it can take little more than the blink of an eye for you to decide that you are interested in someone enough to want to spend the rest of your life with him. If you could only take a little time before making up your mind, you might decide that the object of your interest isn't at all that suitable after all. By the time you get around to such serious thinking, it is way too late. Naturally, sex is a large part of the attraction. When you feel attracted to someone, lust settles like dust over your brain and disengages it. For you, sex must be a shared joy and a large part of the relationship, or the relationship won't work at all.

Aries knows that he/she is the best thing since sliced bread, while Libra likes to talk *at* others, almost as if addressing an audience. You are very fond of your own opinions; you know you are right and that your partner should fall in happily with all your plans and suggestions. This is great if you can surround yourself with mindless people but these would bore you to death. The worst-case scenario for you is to be restricted, to be expected to toe the line, to have to do everything the other person's way and/or to be made to feel powerless. If you choose your friends and lovers from among those who have minds of their own, you can expect a few arguments. If you attract strong and bossy people to you, there will be fireworks. Arguments can also characterise your relationships with family members. The Aries Vertex finds it hard to be diplomatic and to keep its mouth shut – but at least it is honest.

At work, you respect and listen to some bosses and superiors, but you resent those who you consider useless, and you will either fight them openly or try to undermine them. The Libran Anti-Vertex makes you want to be part of a team and the Aries Vertex inclines you towards public service, teaching or the military. Another option would be to work as a self-employed craftsman.

A sixth house Vertex will make your career more important to you on an emotional level than it might otherwise be. In this case, you would put your mind and energies into forging a good career and to reaching an executive position. If you were made redundant or retired, you would feel like a nobody, so you would soon look around for a new way of expressing yourself and expending your abundant energies.

Harrison Ford

Is Harrison Ford the sweet, reasonable man his screen persona portrays? Look at the cardinal Sun, Moon and ascendant in addition to the Vertex and Anti-Vertex. He seems to be a family man, and like all solar Cancerians, he has charm, but he is also tough and self-centred. Interestingly, his other career was as a self-employed carpenter – and apparently he still makes furniture as a relaxation from his film work and he has a reputation for being a real master craftsman.

Rock Hudson

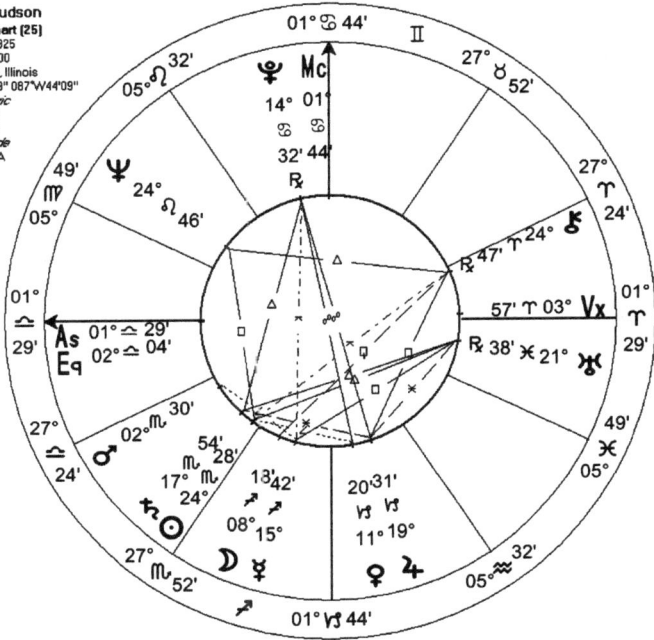

Rock Hudson
Natal Chart (25)
17 Nov 1925
02:15 +6:00
Winnetka, Illinois
42°N06'29" 087°W44'09"
Geocentric
Tropical
Placidus
True Node
Rating: AA

Rock was certainly very handsome, but he never came across as being a sexy man or one who turned-on his female audience, despite being a solar Scorpio. We now know he was gay, so perhaps it was his unadvertised femininity that somehow transmitted itself to us. His Libran Anti-Vertex and ascendant seem to have ensured that Rock came across as a nice guy rather than a sexy one. Aries is a sign that has a good sense of humour, and Rock certainly was amusing in all those light-hearted films. I bet that in relationships, he was not as soft or easy going as his public manner suggested.

Vertex in Taurus

When the Vertex is in Taurus, the Anti-Vertex is in Scorpio. These are both feminine fixed signs. Fixed signs can be obstinate and determined, and the earthy sign of Taurus is notoriously so. Taurus is ultra loyal; it takes a great deal for you to walk away from a commitment, because what you seek is stability.

You don't fall in love in an instant, because you need to allow a certain amount of companionship to grow first. You may be happier with someone from your own background or someone you have known for a while. Foreigners are interesting, but understanding them takes time and too much effort for this Vertex.

Your sensuous nature can make you impulsive at times. While a sudden impulse might lead you into an affair, you need more than a fleeting attraction before you select someone for a long-term relationship. Once you have selected the person you think can fulfil all your needs and desires, things can still go wrong, but it will be some time before you move out or move on. It may also take you a while to realise that relationships are a case of give and take.

The issues that are likely to come up in this relationship are those of control, material goods and money. You or your partner may find it hard to share, or you may both have different ideas about saving and spending. You may disagree over such things as lending, borrowing, mortgages, bills, who pays for what and how much. You display common sense, but you may gravitate to irresponsible partners. Each may try to control the other through finances. There are many possible scenarios here, such as marrying someone who keeps you on a leash or one who sabotages you by wasting every penny that you earn.

I remember one lady with the Vertex in Taurus telling me that her first husband had been a gambler. One day, she came home from work to find that he had sold their furniture and appliances to pay off a gambling debt. Her second husband was a hard worker and a good earner. She and her second husband lived a perfectly sensible family life, spending, saving and sharing money in a reasonable way. It seems that this lady had played out her Vertex karma with her first husband and got her karmic rewards with the second one. There are endless games that people can play with the money ticket – and they don't all have to happen within a marriage.

Naturally, control issues can come in other guises. Some partners are fussy eaters; that's fine as long as they don't make their partners suffer, but with this Vertex, they might just insist on a certain type of cooking. Other areas where one can make the other uncomfortable is if one drinks or smokes and the other doesn't. Here, the non-drinker or non-smoker has a great opportunity to try to control the other.

You must ensure that your lover either shares your priorities in life or that you agree to disagree. Be independent, while allowing your lover his own space. You must also avoid possessiveness and unreasonable jealousy, because that can sour any relationship.

On the other hand, the Taurean gifts of comfort, pleasantness, shared interests in the creation of beauty and sensuality (gardening, cooking, décor, sex) may allow you and your lover to live together in perfect harmony. You may also enjoy harmonious relationships with your family and your many friends. The problems outlined above could be directed towards you rather than perpetrated by you, or they could be a mix of both. Also, second marriages work better than first ones for you.

Princess Anne

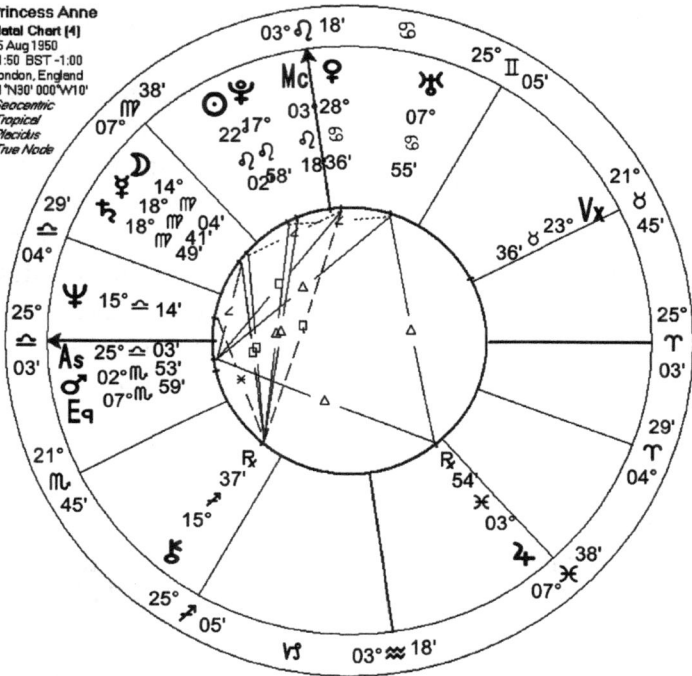

Princess Anne
Natal Chart [4]
15 Aug 1950
11:50 BST -1:00
London, England
51°N30' 000°W10'
Geocentric
Tropical
Placidus
True Node

Have money, possessions and goods been a sticking point in Princess Anne's relationships? Certainly, her royal status and tough, masculine manner must have put a strain on her first marriage. Like most of the Royals, she appears to have been born into the wrong sex, although being a Leo she does seem to have a certain amount of common sense. Perhaps, like Queen Elizabeth I, who Princess Anne resembles, she has "The body of a weak and feeble woman but the heart and stomach of a king!" Her eighth house Vertex suggests resentment and power games.

Billy Connolly

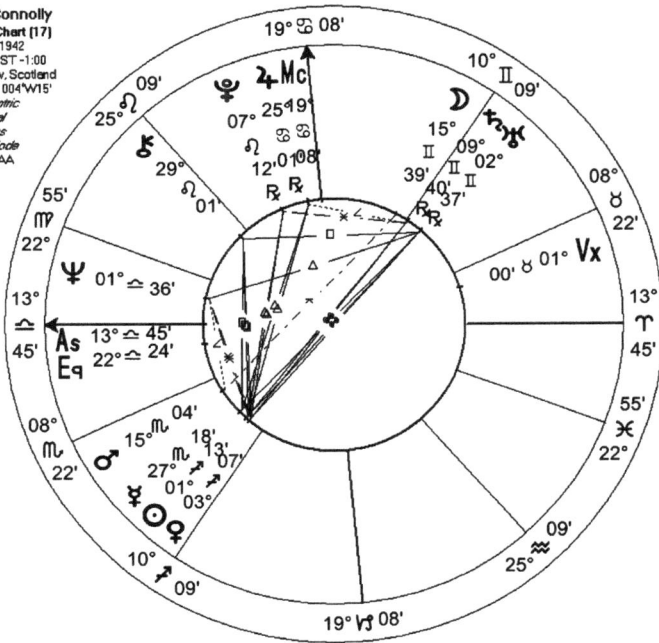

Billy Connolly
Natal Chart [17]
24 Nov 1942
04:30 BST -1:00
Glasgow, Scotland
55°N53' 004°W15'
Geocentric
Tropical
Placidus
True Node
Rating: AA

One would not think that finances would matter to a cheerful Sagittarian such as Billy. However, money represents freedom and to someone who came from the wrong side of the tracks, that is what it would provide. Billy comes across as a thoughtful, spiritual man and his Taurus Vertex helps him to value his partner and to keep both of them happy. It now emerges that his father subjected Billy to horrendous sexual abuse throughout his childhood. That must truly be the evil face of the Scorpio Anti-Vertex!

Vertex in Gemini

When the Vertex is in Gemini, the Anti-Vertex is in Sagittarius. These are both masculine mutable signs, which means that you need variety in your life. Gemini is an air sign and Sagittarius is a fire sign.

Although far from being stupid in other spheres of your life, when it comes to falling in love, you can do so at the drop of a hat and with someone who is wrong for you. It seems that you have to learn your karmic lessons the hard way. You must be able to talk with your partner and to share ideas. His mind must be as quick as yours or you will soon become bored with him. The problem here is that you over-analyse things and you can work yourself up over nothing.

You are drawn to people who you can talk to, listen to and debate with. You are fascinated by ideas and concepts that are new to you, so your friends and lovers will open your mind to new ideas. You may meet your partner in an educational arena or you might spend time teaching others once you have got together. You cannot fancy a dimwit, however handsome and charming, because the way that a person's mind works is as important to you as their looks or character. Having said that, this is not a sexless Vertex position, because sex is another form of communication that you wish to explore and enjoy. There is a certain tension in your nature that makes sex a natural and relaxing outlet for your nerves, and you have an explorative turn of mind where lovemaking is concerned.

Friendship is as important to you as love relationships, so you will have many important friends. You may compartmentalise your friends, so that there are those with whom you go to the ballet or opera, and those with whom you go to astrology lessons or meet in the pub. One thing your friends must have is a sense of humour, because you like to laugh.

Your worst relationships are likely to be those that you had early in life. Your family may have little time for you and you may have had good reason to loathe your schoolteachers or classmates. On the other hand, you may have loved every moment of your school days, because school and the various associated activities (sport, drama, computer club) may have offered you a legitimate escape from a critical and unloving family.

An unhappy childhood leaves anyone with a poor self-image and this can play itself out as a tendency to select selfish, sarcastic, thoughtless or violent partners. This is the last thing your fragile ego needs. On the other hand, you may be rather dictatorial yourself, with a tendency to forget that your lover is entitled to his own ideas, methods, opinions and life.

Gemini likes to work, so this Vertex position ensures that you will always have a job to do, and many of your friendships will be formed through your work. The emphasis on work and career will be all the stronger if your Vertex is in or near the sixth house.

Germaine Greer

Germaine Greer
Female Chart [23]
29 Jan 1939
06:00 AEST -10:00
Melbourne VIC, Australia
37°S49' 144°E58'
Geocentric
Tropical
Placidus
True Node
Rating: A

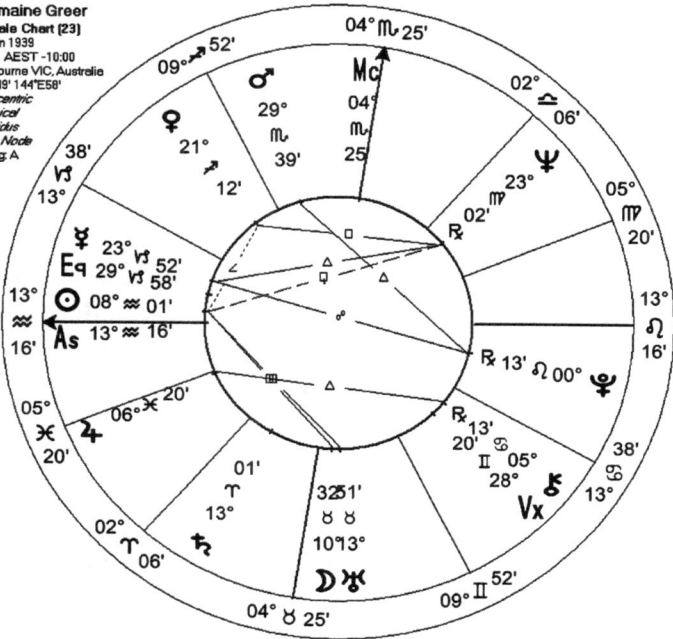

This lady has been a great writer, critic and communicator for over thirty years; she must be fascinating to listen to and learn from, but could one live with her? All I have ever seen and heard her do is criticise everything and everyone. She only seems to revere the place that she lives in – probably because she lives there!

Her Gemini Vertex accounts for her career in journalism and TV, in addition to her teaching and writing work.

Uri Geller

Uri Geller
Natal Chart [34]
20 Dec 1946
02:00 -2:00
Tel Aviv, Palestine
32°N04' 034°E46'
Geocentric
Tropical
Placidus
True Node
Rating: A

Uri seems to have a reasonably happy family life, and he is probably great fun as a friend; he is certainly loyal, as witnessed by the way he stood behind Michael Jackson all the way through Michael's traumatic court case. Like many Sagittarians, he is enthusiastic about spiritual life and he is also very sporty. He comes across as a real nut at times, but he is sincere about what he is and he is definitely a great communicator. The Gemini Vertex in the intense eighth house shows a fear of abandonment. Is this perhaps a hang-over from a previous life experience?

Vertex in Cancer

When the Vertex is in Cancer, the Anti-Vertex is in Capricorn. These are both feminine cardinal signs, although Cancer is a water sign while Capricorn is an earth sign. Cardinal signs like to have their own way and they aren't great at compromising. They barge ahead, often convinced that they know what is best for themselves and for others.

You do not fall in love particularly quickly, but others seem to fall for you in an instant. Whether they truly love you for yourself or for the home, money, family life, help with their work or even the amusement that you appear to offer them, is a moot point. As it happens, you would provide a lover with a good home and a listening ear when he wants it, but the price of these things might be very high. You demand far too much of your lovers, your family and even your children. You can be sarcastic and hurtful when they don't come up to your impossible standards or when they don't put your needs centre stage. You pour your love into your family but you want it returned by the ocean-load, and nobody can do this to your satisfaction, so you may end up driving away the love that you so badly need.

Some astrology traditions link Cancer with the mother or motherhood and Capricorn with the father or fatherhood, so there is likely to be a strong possibility for parent and child games to be played out in a relationship. You may choose to mother your lovers or have them play the part of the parent. This is fine as long as you and your lover are happy with the arrangement, but trouble will come if you take this too far and monitor your partner's diet, wardrobe, choice of friends and so on. You have a wonderful capacity for love and for nurturing others, so if you can tap into this and stop trying to dominate them and stop making impossible demands, you could end up with the best love relationships of the zodiac.

You may be on the receiving end of poor relaters rather than being a poor relater yourself. For instance, your parents or your partner's parents may make your life difficult. They may be demanding, critical or just too involved in your life. You and your lover may never be able to take a holiday or go to a party without having to take one or more relatives along. On the other hand, you could enjoy being part of a close family, and that might include your partner's family as well as your own. You could all look out for each other and help each other in a responsible, kind-hearted, humourous and loving manner. You could be extremely close to your mother.

Other problems might surround money (as a source of power) or business. You might switch from being cautious and sensible one day and taking decisions on an emotional basis on the next. There may be disputes about work, especially if you or another family member work in or around the home – in market gardening or farming, perhaps.

This Vertex is patriotic and fond of its homeland, so it can take you into politics.

Shirley Temple Black

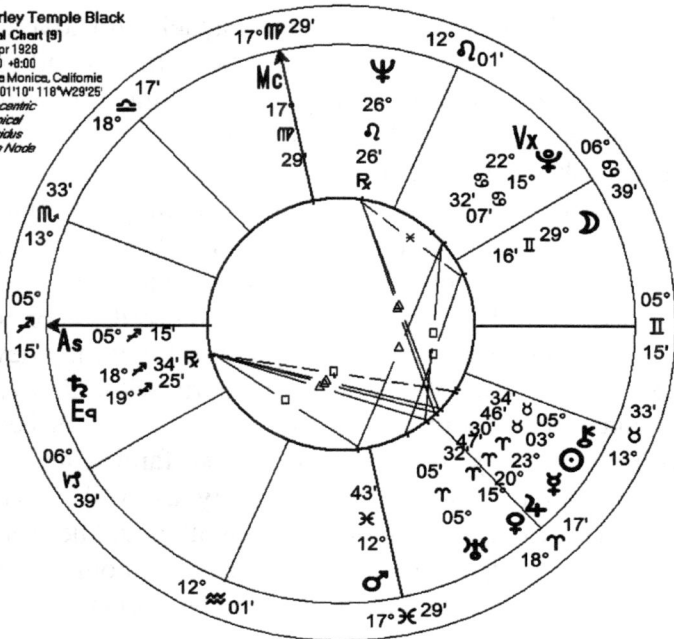

Shirley Temple Black
Natal Chart (9)
23 Apr 1928
21:00 +8:00
Santa Monica, California
34°N01'10" 118°W29'25"
Geocentric
Tropical
Placidus
True Node

Well, if there was ever a case of a pushy show-business mother making her child a star, it was Shirley Temple's mum. In a TV interview, Shirley said that she enjoyed the experience, but the eighth house Vertex says that she might also have been somewhat resentful about it. One could say that Shirley turned her audiences into parental or aunt & uncle figures.

Interestingly, as an adult, Shirley went into politics.

Marlon Brando

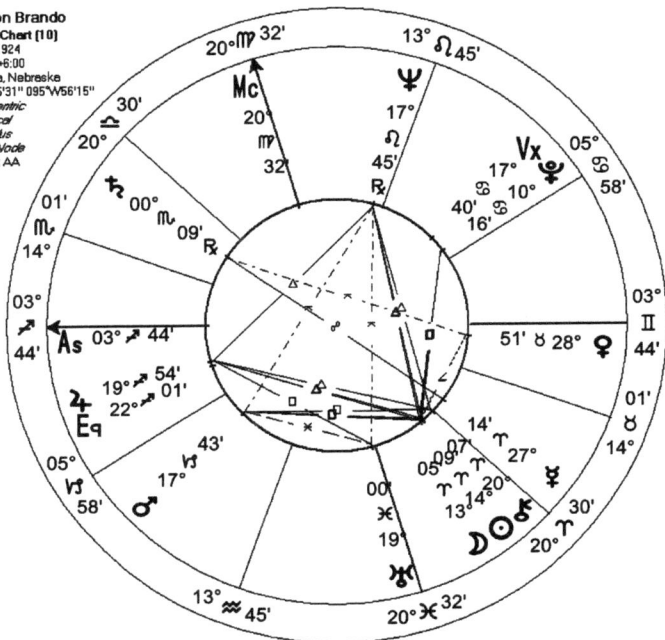

Marlon Brando
Natel Chart (10)
3 Apr 1924
23:00 +6:00
Omaha, Nebraska
41°N15'31" 095°W56'15"
Geocentric
Tropical
Placidus
True Node
Rating: AA

Moody, magnificent Marlon is a bit of a mystery, but we do know that one of his marriages was to a Native American woman and that he was extremely interested in the Native Americans. Perhaps in some way, they became his family, and he certainly supported their aims in the political arena. Interestingly, he had little interest in his Italian roots and almost turned down the chance of appearing in "The Godfather" because of that. It appears that Cancer relates more to mothers than fathers – even godfathers.

Vertex in Leo

When the Vertex is in Leo, the Anti-Vertex is in Aquarius. Both are masculine fixed signs, which denote obstinacy, determination and power. However, Leo is a fire sign while Aquarius is an air sign, so the inclination is to jump first and think about what you have done later.

Leo is a romantic sign, so you fall in love in an instant, and sometimes it is lust and sexual heat that drive your emotions. This fiery, sexy sign leads you to choose your lovers fairly blindly, and then when the dust of lust starts to settle, you wonder what you saw in them. The one thing that saves you from too much agony is that you get bored with your paramours quickly, so you find it relatively easy to dump them and move on. Where longer-term relationships are concerned, the Aquarian side kicks in and urges you to chose those who you can like as well as fancy. Then you can enjoy talking with them, respecting them and sharing values and interests with them.

Leo is a misunderstood sign. Many astrologers think that Leo likes to be the centre of everyone's attention and to be centre stage. This is not so. Leo has a great deal of ability and common sense, so it naturally gravitates to being in charge and making the decisions on behalf of others. Fortunately, this sign has everybody's interests at heart and it makes a benign boss. However, this means that you attract lame ducks, both within your family and among your friends and as lovers. They are attracted to your strength. You can tip over into being domineering and dictatorial though, so you must watch this.

Issues are of total independence against having a crowd of people depending on you. Your sense of responsibility might become your downfall, as you can wear yourself out on behalf of work associates, family, friends, clubs, groups, societies, humanity in general or your partner.

Children might be an issue, because while Leo wants to have them and to put them first, Aquarius is happiest without them. You might have mixed feelings about children or you might attract partners who have mixed feelings about them. You may also attract childish partners or tend to be treated like a child yourself.

As Vertex signs go, this is not a bad one, as Leo is a very affectionate, generous and loving sign and even your Anti-Vertex can cope with family life as long as there is real love there. The Leo Vertex means that you are loyal and loving towards relations and friends. If a friend lets you down, you become so shocked and upset that it can take you a long time to recover. Though you may be polite to the offending person when you meet them in future, the trust, affection, friendship and esteem that you previously had for them will have gone forever.

In addition to a busy life, you may also care for pets and small animals.

Work-wise, you will put much energy and effort into your career, thus being very productive.

Jamie Lee Curtis

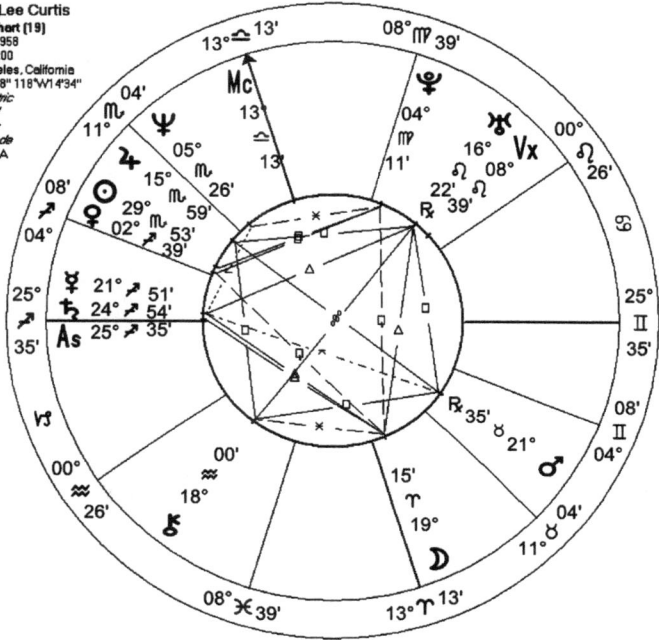

Jamie Lee Curtis
Natal Chart (19)
22 Nov 1958
08:37 +8:00
Los Angeles, California
34°N03'08" 118°W14'34"
Geocentric
Tropical
Placidus
True Node
Rating: AA

Older forms of astrology associate Leo with the father and it is well known that Jamie is very fond of her father, although his alcoholism must have made him a pain to live with at times– as reflected by the resentful eighth house position of her Vertex. Jamie comes across as a nice person, and she must be great fun to be with.

Lucille Ball

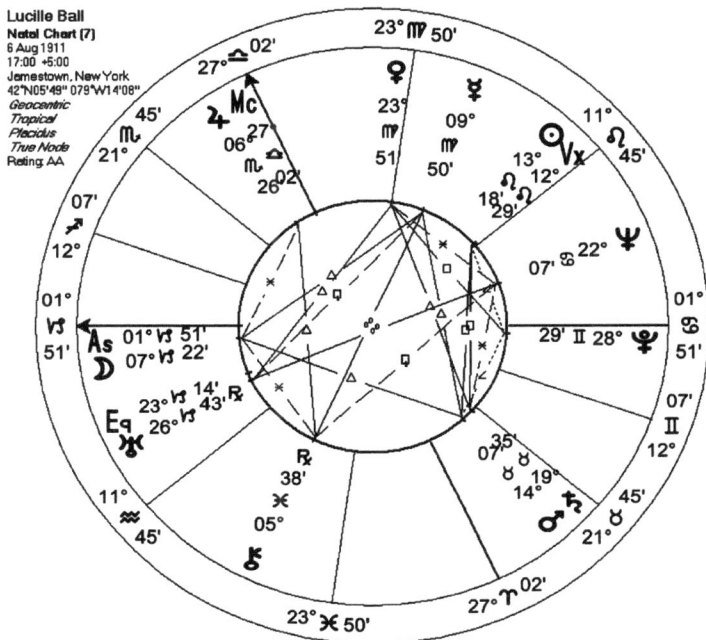

Lucille Ball
Natal Chart (7)
6 Aug 1911
17:00 +5:00
Jamestown, New York
42°N05'49" 079°W14'06"
Geocentric
Tropical
Placidus
True Node
Rating: AA

It is well known that Lucille Ball was a capable businesswoman who literally ran the show. Not surprising with her Capricorn ascendant and Moon and her eighth house Sun, but look at the conjunction to the Vertex. She rose to fame with her husband Desi Arnez, until his womanising finally got her down enough to call a halt to their partnership. Lucy didn't seem to bother too much with marriage after Desi, so perhaps her independent Aquarian Anti-Vertex took over, allowing her to have more fun with her many friends than she had within her marriage.

Vertex in Virgo

When the Vertex is in Virgo, the Anti-Vertex is in Pisces. Both are feminine, mutable signs denoting a need for variety in life, but Virgo is an earth sign while Pisces is a water sign.

You are not noted for falling head over heels in love in an instant although you can do so on occasion. Once you decide that you love someone, you don't lose interest in them too quickly, and even though you may part company, affection and friendship often linger on afterwards. Casual affairs are a different matter though, because you can sleep with a number of people who you can hardly remember a week after the event.

I consider Virgo to be the sign of the donkey! You know the kind I mean… it has panniers on either side of its back, and they are full of other people's stuff. This is such an apt picture of Virgo carrying everyone's load in addition to his own.

Whenever a friend or relative asks you to give a hand at a party, you are there in an instant with your apron on. If someone needs to be cared for, you are the one who will do it. You want others to like you and to approve of you, so you give them your time, energy, money and goods. You enjoy feeling involved in whatever is going on around you and being in control of events to some extent. You may choose a partner who is ailing, alcoholic or in some other way dependent on you. Alternatively, your partner might be fine but you fill the house up with children (not all yours), grandchildren, relatives and hangers-on. You may work in the field of health and healing, charity fund raising or social work. You genuinely want to help the disadvantaged and those whose bodies or souls are in pain.

You loathe it when others criticise you – partly because you might have suffered from too much criticism when you were a child. However, you do sometimes give yourself permission to criticise others, and this may become a bad habit as you get older. Nobody will want to be friends with someone who can only see the worst in others. Nobody's perfect, not even you – but we all have good in us, and that is what you should try to focus on.

You like to be busy, so you are likely to have a regular job or career. You don't like chopping and changing, so you can stay in the same job for quite a while. If your Vertex is in the sixth house, your career will be extremely important to you. There is a strong possibility of working in writing, journalism, broadcasting, or in the fields of health and healing.

Mark Knopfler

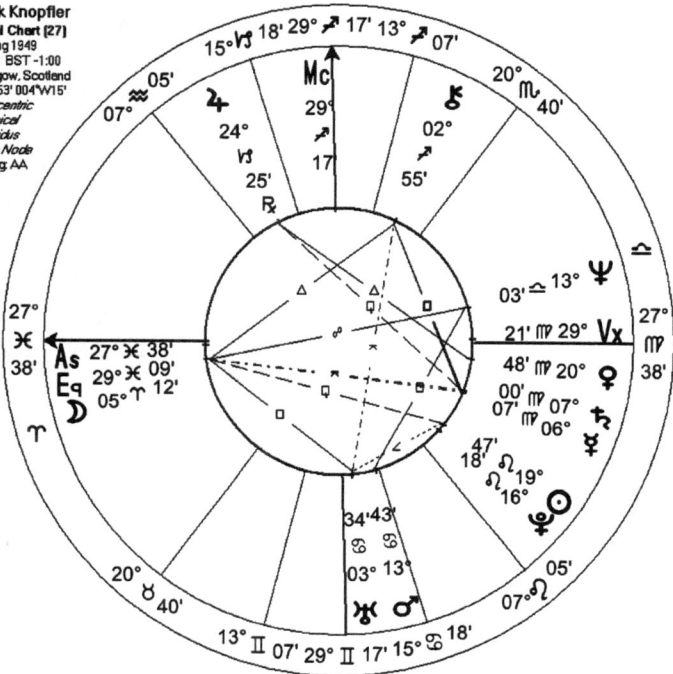

Mark Knopfler
Natal Chart (27)
12 Aug 1949
21:50 BST -1:00
Glasgow, Scotland
55°N53' 004°W15'
Geocentric
Tropical
Placidus
True Node
Rating: AA

We don't know much about Mark as a person, but we admire his musicianship and the songs he sang with Dire Straits. He is no looker – but his wasted appearance seems to translate into a strange kind of elegance. Do Mark's Pisces ascendant and Anti-Vertex make him sacrifice too much for the sake of others? Probably.

Larry Hagman

Larry's quirky Aquarian ascendant and Anti-Vertex were admirably displayed when he acted in the TV series, *I Dream of Jeanie*. In that program, he put himself out time and again for his strange, witchy wife. Larry's Virgo Vertex makes him a hard worker. It is likely that his relationships are with those he meets through his work or perhaps through voluntary jobs.

Vertex in Libra

When the Vertex is in Libra, the Anti-Vertex is in Aries. Both are masculine, cardinal signs, which means that in relationships, you want to do things your way. Libra is an air sign and Aries is a fire sign, which means that the desire to think things over before acting battles with the urge to push ahead with plans at full speed.

Your love style is confusing – even to you! You can fall in love in an instant but then wonder if you are doing the right thing. You probably have the highest level of sexuality of all the Vertex types but there is a monumental battle that goes on between your heart and your mind (and even your mouth) as you talk yourself out of a relationship as quickly as you fell into one.

The Vertex is comfortable in Libra because the Vertex is concerned with relationships, as is the sign of Libra, so there is no reason why you shouldn't be able to fall in love, marry, have good friends and enjoy family life. You have considerable charm, which encourages others to respond to you with warmth, and your luck in the relationship areas of life should be better then most. There is a sexy air about you, and you may indeed have a high libido. You certainly can be flirtatious and very attractive, so you will never find it hard to interest others, and this may lead you into a series of affairs. One might think that the laws of karma mean that you will be punished for your behaviour by ending up alone and embittered, but this fate is unlikely. Despite your naughtiness (or perhaps because of it), you manage to keep the affection of your parents, siblings, children and friends.

Both Aries and Libra need lovers, partners and families, but Aries wants to make decisions for himself and for his family, while Libra finds it hard to be decisive. This makes it hard for others to know what you actually want sometimes. You can drift away from situations that get on your nerves, so your lovers are never quite sure of you. Perhaps this adds mystery to your charm? Your family relationships should be fine, but the Anti-Vertex in Aries means that your parents, siblings or even children insist on having their own way at your expense. If their demands become unbearable, you will walk away from them all.

The Libra Vertex can make you argumentative and confrontational and the Aries Anti-Vertex makes this inevitable. These are cardinal signs, which means there is a strong desire to do things your way. If you feel that you must win every battle in your relationship, you could end up losing the war. However, you have a strong sense of justice, so perhaps you could try to apply this to those who you sometimes feel urged to dominate and bully.

You may be interested in music, the arts, acting, performing or being on show. Alternatively, you may be attracted to people who do those things and you certainly want to be among fascinating and amusing people. However, the most impressive aspect of this combination is its warm, wonderful, seductive sexual overtone that transmits itself so easily to others...

Christine Keeler

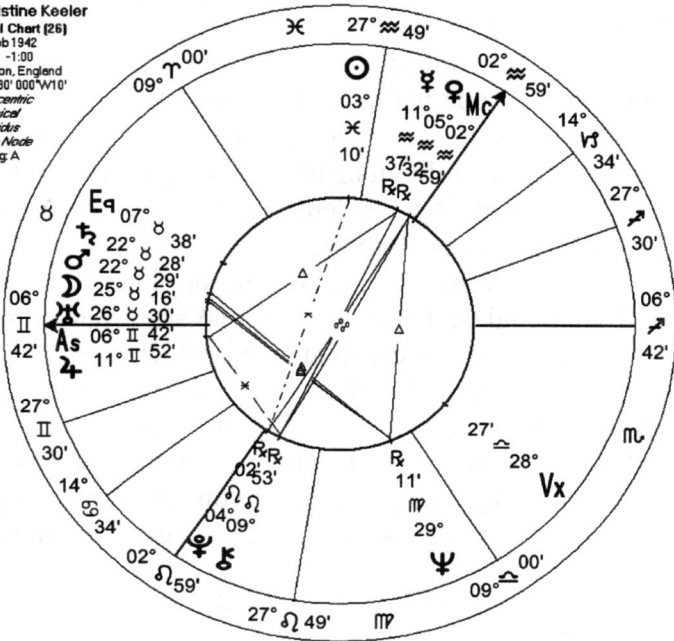

Christine Keeler
Natal Chart (26)
22 Feb 1942
11:15 −1:00
London, England
51°N30' 000°W10'
Geocentric
Tropical
Placidus
True Node
Rating: A

In the 1960s, the beautiful callgirl, Christine Keeler, and "happily married" Minister for War, John Profumo, were caught having an affair. Unfortunately, Christine was also sharing her favours with a Soviet spy – in the middle of the Cold War!

Christine's Libran Vertex certainly gave her charm and looks. The dual nature of Libra (Scales), along with dual Gemini rising, (Twins) and her Sun in dual Pisces (Fish) show that more than one love in her life at any one time was inevitable for this lady. Her Moon in Taurus gives her a luxury loving nature along with the beauty and sexuality of her Libran Vertex. The fact that her Vertex is in the sixth house shows that to her, sex and work were one and the same thing.

Barbara Cartland

Barbara Cartland
Natal Chart (2)
9 Jul 1901
23:40 +0:00
Edgebaston
52°N30' 001°W50'
Geocentric
Tropical
Placidus
True Node
Rating: AA

Barbara Cartland also had the Vertex in Libra, in the sixth house. She was very pretty when she was young and she had many admirers. Had she been born a couple of decades or so later, she would probably have had lots of lovers, but in the absence of these, she poured her interest in romantic love into 400-plus books instead. Here again, love and sexuality became part of this lady's work.

Vertex in Scorpio

When the Vertex is in Scorpio, the Anti-Vertex is in Taurus. Both are feminine fixed signs, although Scorpio, being a water sign, is more flexible than the obstinate earth sign of Taurus.

You can fall in love at a glance, possibly due to sexual attraction or due to emotional vulnerability. Even if you choose to love the wrong person, it takes you a great deal of time before you give up any hope of a future. When love ends, it takes a long time before your sensitive heart can heal. In a way, you never really do get over anything or anyone, so you carry around a heavy bag of negativity. Your worst problem is that you can choose to love someone who hasn't much going for him, and then put him on a pedestal. You can mentally endow a pig's ear with a silk purse level of nobility, intelligence, integrity, understanding and value.

There is a level of intensity here that can take a variety of routes. Relationships may bring you heights of love, passion, affection, sexual fulfilment and a sense of deep commitment. You seek out deep thinkers and those who are in positions of power and authority, but these people may never really understand you or give you the support and affection that you need. This Vertex will give you a lifetime of highs and lows. One of the reasons for your attraction to strong, powerful people is that you feel yourself to be weak and powerless. Perhaps you were put down in childhood by a powerful parent who made you feel insignificant. Somehow, you have grown up thinking that the opinions of others are more valid than yours and that they will make the right decisions on your behalf. Alternately, you continue to fight against your rough childhood throughout your adult life by resenting your loved ones and by becoming impossible to live with.

You may fall in love with a married man who gives you some blarney about how he will definitely leave the wife when his youngest child reaches 16, leaves college or collects its old age pension. You may start out with parents who dominate and suppress you, then move on to partners who do the same. Your so-called loved ones might criticise your appearance, behaviour, clothes, friends, job and much else. They may try to control what you eat, drink, smoke and do with your life. This can be so counter-productive that you end up eating, drinking and smoking yourself into an early grave just to spite them. Try to ensure that your choices in life are just that - *your* choices, not those forced on you by others.

It may be impossible for you to find love and sex in the same place, so you may have a lover who is great in bed and another who keeps you happy by just being with you. With luck, you may eventually find someone mentally, emotionally and spiritually there for you, and with whom you are very happy.

Not all of your relationships are intimate, so you may have some extremely good ones with parents or colleagues. You are a good friend to others because you can empathise with their problems and be a shoulder for them to cry on.

Your Anti-Vertex is in Taurus, which rules personal values and priorities, money, goods, resources and wealth. Money, wealth and goods are frequently controlling factors in any relationship, so there is a chance that your partner will play games with you, keep you under control, use you, lean on you, keep you short or be unfair to you in this area of life.

Singing, dancing or acting may be an important part of your life.

Patsy Cline

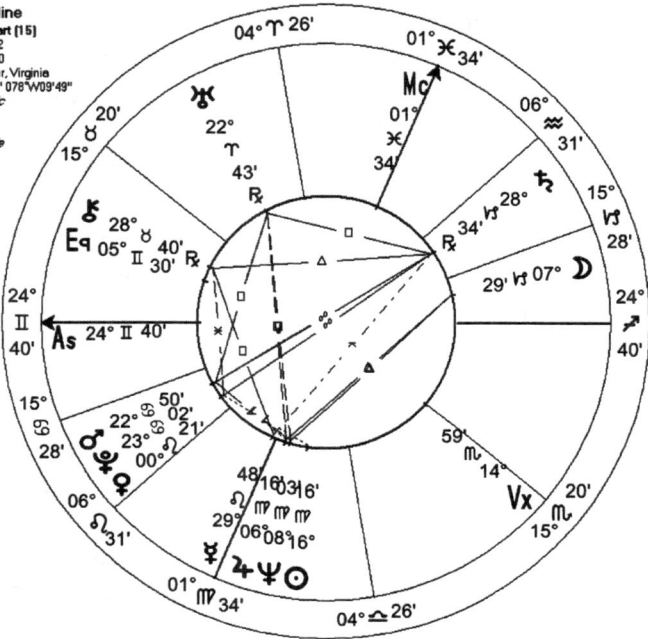

Patsy Cline
Natal Chart [15]
8 Sep 1932
23:15 +5:00
Winchester, Virginia
39°N11'08" 076°W09'49"
Geocentric
Tropical
Placidus
True Node
Rating: AA

Younger readers won't know this lady, but she wrote and sang songs that became classics in their time, such as "I'm Sorry" and many others. Patsy was a hard working country singer whose childhood, marriages and love affairs were extremely destructive. She died young, ostensibly of ill health but possibly through drugs, suicide or something of the kind. Her Vertex is in Scorpio on the sensitive fifth/sixth house cusp.

Sam Cooke

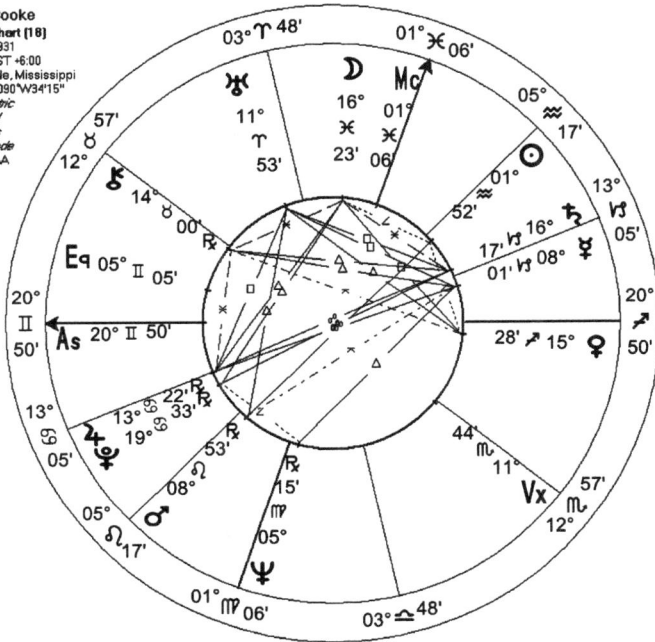

Sam Cooke
Natal Chart [18]
22 Jan 1931
14:10 CST +6:00
Clarksdale, Mississippi
34°N12' 090°W34'15"
Geocentric
Tropical
Placidus
True Node
Rating: AA

This handsome, sexy singer and writer introduced many of us to soul music of the type that will never be forgotten (*I Heard It On The Grapevine*). It is probable that many children have been conceived while his music was playing nearby. Sam had some kind of on-going problem with his father, who may have been an alcoholic or druggy. Sam's life came to a premature end when his father shot him during an argument. Interestingly, Sam's Vertex is in exactly the same position as Patsy Cline's.

Karen Carpenter

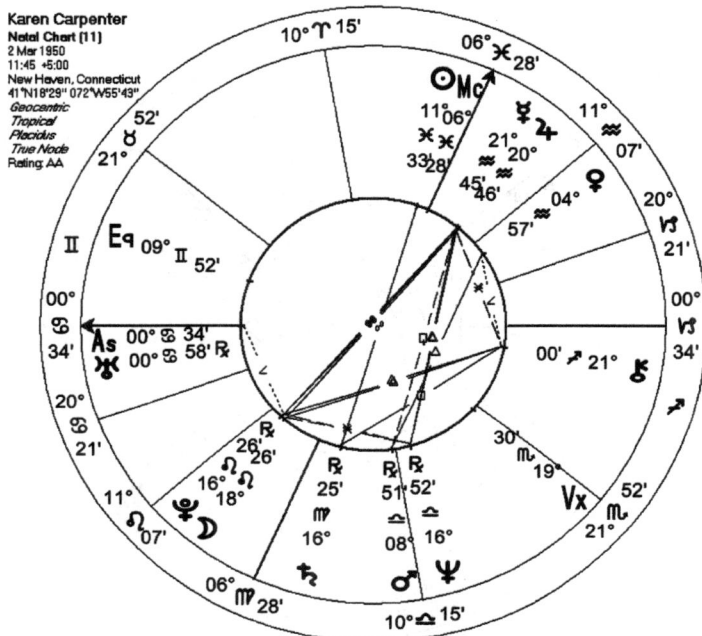

Karen Carpenter
Natal Chart [11]
2 Mar 1950
11:45 +5:00
New Haven, Connecticut
41°N18'29" 072°W55'43"
Geocentric
Tropical
Placidus
True Node
Rating: AA

I have to add a third chart here because the coincidence is so powerful. The Vertex is once again in Scorpio, on that same sensitive cusp. Karen Carpenter was yet another singer-songwriter. She was close to her brother who played and wrote much of their wonderfully romantic music. She was extremely fragile mentally and emotionally, and when her short-lived marriage failed, she starved and exercised herself to death.

Vertex in Sagittarius

When the Vertex is in Sagittarius, the Anti-Vertex is in Gemini. Both are masculine mutable signs, but Sagittarius is a fire sign while Gemini belongs to the element of air. Being a fiery sign, the Sagittarian Vertex means that you fall in love very quickly, but you can be silly and unrealistic, so you may choose the wrong kind of partner on occasion.

You may be highly sexed and inclined to experiment. Others see you as a person who bubbles with energy and enthusiasm and who has many interests. You may have a number of short-term lovers who attract you at first, but soon bore you after a period of experimentation. You can meet the right person and settle down, but they must share your interests and be a friend as well as a lover. Whether or not you remain faithful after that depends upon many other factors, including your partner's attitude.

Some of you may be much better at friendship than commitment. You like short-term friendships that you can enjoy for a while and then leave behind, while you move on, making new friends.

It is likely that you lean towards religion, philosophy, Wicca or any mind, body and spirit subject. Materialistic people bore you, so you seek to share your life with those who are more interested in the spiritual side of life than in making money. You seek out those who are as idealistic as you are. It seems to me that you will probably have a happy love life, although you may accumulate little in the way of material goods, unless there is something on your chart that makes you money minded. You and your lover may become involved in teaching or in producing newsletters or websites devoted to the subjects that fascinate you.

Travel will feature in your life but you may be more interested in spending time in an Ashram, trekking in mountains or jungles or working as a volunteer in a third world location than wallowing in five-star luxury. It is entirely possible that some of your lovers will tend to be foreigners.

There is a kind of innocence about you that attracts down-to-earth types, so you may be drawn to motherly or fatherly people. You may be fond of your parents and siblings, but chances are that you will have long since gone your own way. Nor will you seek to hang on to your own children once they have grown up.

One fault causes you more problems than it does to others, and that is your lack of judgement and a lack of realism. Another fault is that you can be cutting, sarcastic and hurtful - and that can cost you dear in terms of love and friendship.

Something that can confuse your partners is your need for independence. This means that you want to love and be loved, but not to be smothered or controlled.

Jeffrey Archer

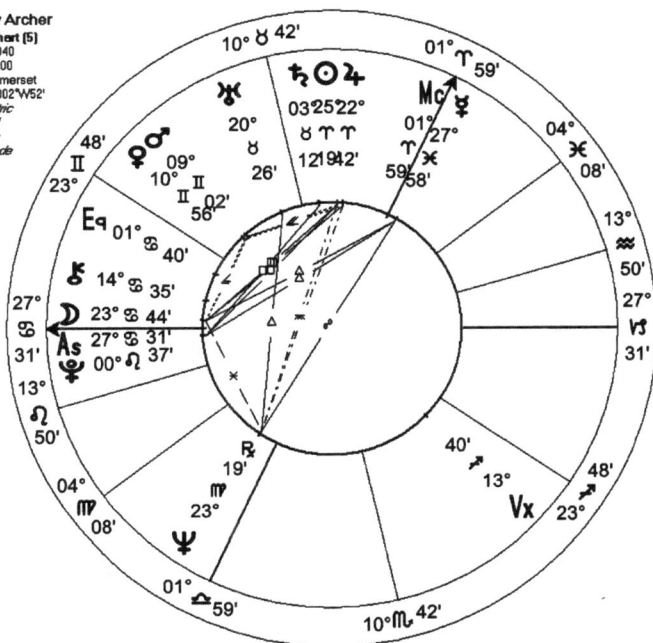

Jeffrey Archer
Natal Chart (5)
15 Apr 1940
11:45 -1:00
Mark, Somerset
51°N00' 002°W52'
Geocentric
Tropical
Placidus
True Node

Love him or hate him, Jeffrey certainly made a stir. Like many with the Sun in Aries, Jeffrey started out as an idealist, but his idealism was mixed with a lust for power. Interestingly, his wife the "fragrant" Mary, works in a humanitarian field. Being Cancerian, it is possible that she was not above perjuring herself in her man's defence. The Sagittarian Vertex can lead to sexual indiscretions, especially when in the fifth house. Poor Jeffrey is very bright but he lacks judgement, so indiscretions and allying himself to the wrong people have caused him many serious problems.

Edwina Currie

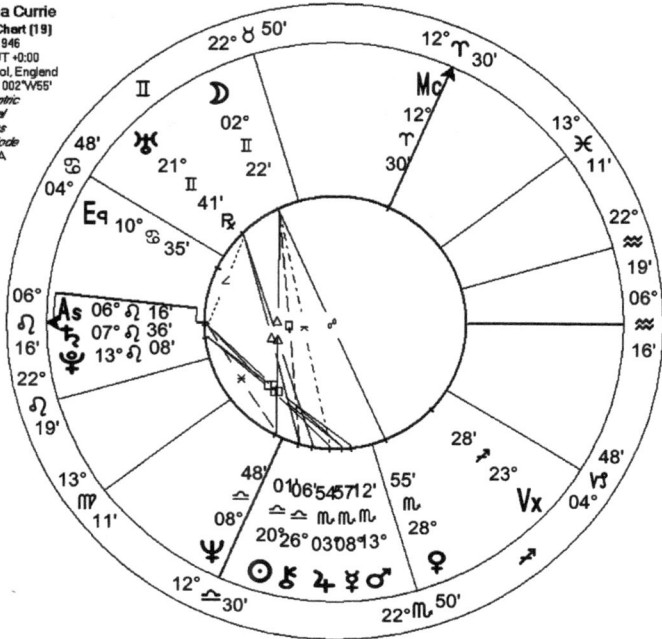

Edwina Currie
Natal Chart [19]
13 Oct 1946
23:30 UT +0:00
Liverpool, England
53°N25' 002°W55'
Geocentric
Tropical
Placidus
True Node
Rating: A

Attractive, talented Edwina has turned herself into a successful author and a television personality, but her first love was politics and her driving force was idealism, as shown by her Vertex in Sagittarius. As an MP, it is interesting that she attached herself to John Major, a man who was going places, but who dumped her after he got there. The lack of judgement and realism in the Sagittarian Vertex shows itself here as well. Edwina was one of the few women to make it into the cabinet, but others ensured that she didn't stay there long.

Incidentally, Piscean Paddy Ashdown has the same Vertex placement, and he, too, had an affair that became public.

Vertex in Capricorn

When the Vertex is in Capricorn, the Anti-Vertex is in Cancer. These are both feminine cardinal signs. Capricorn is an earth sign while Cancer is a water sign. When researching this Vertex sign, I came across one serious, hardworking and successful person after another and it seems that hard work is the guiding principle in the person's life.

Falling in love means temporarily giving up any semblance of common sense or even sanity, and that doesn't sit easily with the sign of Capricorn. When you fall in love, you really mean it, but this can bring problems. The good news is that your youthful romances may not last, but your later ones do. Perhaps this is because you learn by experience and make better choices the next time around.

If this is your Vertex sign, your job or serious interests will fill your life. Your partner will either have to put up with the fact that you are at work most of the time, or work alongside you. This is quite possible, because you tend to meet your friends and lovers at or through work. Indeed, you may feel uncomfortable when among those who are not in your field or who do not operate at your level. Capricorn anywhere on a chart shows ambition and a slow climb up the ladder of success. When the Vertex is in this sign, friends, partners and relatives must all understand this or move on to someone who is more available.

There may be something in your background or family history that makes you such a hard and competent worker. Somehow, the message that you must make it at all costs seems to have set in when you were very young. In adulthood, you may take on the responsibility of caring for parents or other relatives, either hands-on, or by providing money for them.

Your friendliness and sense of humour attract others and are certainly things that you look for in others. You may be interested in travel, so your partner should also enjoy seeing the world. Your fondness for details can take you into the world of hallmarks, stamps and coins, collecting old records or antiques as hobbies.

The main thing is that you need a partner who sees the importance of being dutiful, thorough and respectable. Your partner should have an acceptable religion or philosophy, and fit into society without making a fool of himself. Even if you are not ambitious, careful, thrifty, sensible and responsible, you may attract the type of person to you who is.

You must guard against putting your career before your feelings and emotions. You must try to create a balance between your need for power and your need to give love – and not just money – to your family, friends and lovers. Otherwise, you may end up impressing others but never being close to anything other than your bank balance.

Muhammad Ali

Nobody can deny that Muhammad Ali worked hard to become the world's most memorable and charismatic heavyweight champion. His Leo ascendant gave him his outward glitter and cheekiness, but the Sun and Vertex tell a story of many years of work, work and more work.

Like many black people of his generation, Muhammad used his talent and intelligence to pull himself up from the poverty of his birthright.

Natalie Cole

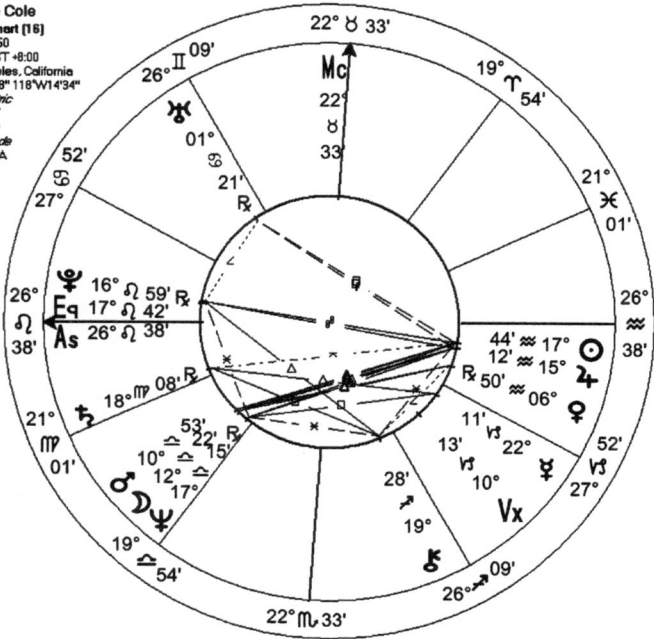

Natalie Cole
Natal Chart (16)
6 Feb 1950
18:07 PST +8:00
Los Angeles, California
34°N03'08" 118°W14'34"
Geocentric
Tropical
Placidus
True Node
Rating: AA

Here we see the standard interpretation of a Capricorn Vertex - following in father's footsteps. The lovely Natalie has carved out her own career, but her dad must be smiling down proudly at her from the world of spirit, and certainly gave her the right start to her career.

As expected with this Vertex, Natalie is an attractive and very hard-working lady.

Vertex in Aquarius

When the Vertex is in Aquarius, the Anti-Vertex is in Leo. These are fixed signs that denote a certain amount of determination and obstinacy. Aquarius is an air sign while Leo is a fire sign.

If this is your Vertex sign, the chances are that friendships take precedence over close family relationships, marriages or even lovers. Even within love or family relationships, there has to be a great deal of friendship for the relationship to function or to survive. You may long to be close to others, but somehow maintain a distance from them that doesn't allow relationships to flourish.

The Leo end of the equation suggests that relatives and lovers might be in the entertainment business or something equally glamourous or they might encourage your interests in this direction. There is a search for charisma, respect, fame, nobility and notability going on around you or perhaps within you. You may be slower to fall in love with others than they are to fall for you. Indeed, you may never quite get the hang of abandoning yourself enough for you to be able to fall deeply in love. Confusingly, the Aquarius Vertex can make you overly possessive.

You may have an unusually close relationship with a parent or you may play the role of parent or child in your relationships. It is likely that you will have children and you would either have a wonderful relationship with them or a dreadful one. You could be a disastrous parent and you may eventually detach from your offspring and end up feeling indifferent about them. On the other hand, the parent child relationship may be the one that really works for you. Everything about this Vertex combination could be highly successful or a complete washout.

You certainly have charisma and it might be your intelligence, your notoriety, your talent for comedy or your "different-ness" that fascinate others. You have strong political or humanitarian views and a potential partner must understand this side of your nature. You may explore bisexuality or even a mixture of sexualities as you go through life. Naturally, all this strangeness may be heightened or muted by other factors on your horoscope.

There is a conflict between your strong sense of duty and commitment and your need to be completely free to operate as you wish. You are dutiful but detached. You are hard to understand and hard to live with but you can make a wonderful friend – and sometimes, along with a few sexual flings here and there – that is enough for you, and for those who surround you.

Marlene Dietrich

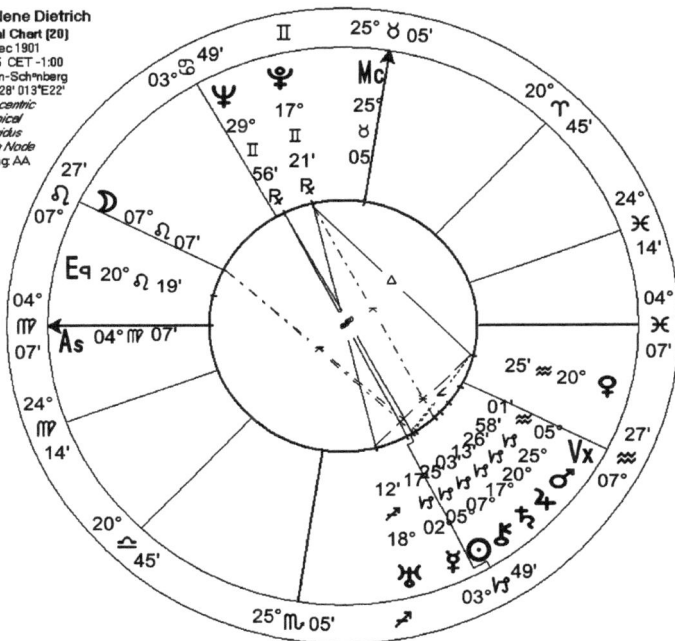

Marlene Dietrich
Natal Chart (20)
27 Dec 1901
21:15 CET -1:00
Berlin-Schönberg
52°N28' 013°E22'
Geocentric
Tropical
Placidus
True Node
Rating: AA

What a hard-working woman this was! Her attention to detail was phenomenal. We know that acting, singing, dancing and entertaining were her forte, but a slight difference in her chart might have made property development her winning formula. Not much was said about such things in her day but she was known to be bisexual. She had affairs with all her leading men and sometimes her women as well. She was sexually voracious, probably impossible to live with, but she was sooo endlessly fascinating!

Peter Sellers

Peter Sellers
Natal Chart [31]
8 Sep 1925
06:00 BST -1:00
Southsea, Portsmouth
50°N46' 001°W05'
Geocentric
Tropical
Placidus
True Node
Rating: A

We all loved Peter for his ability to act and to imitate
and for his wonderful Virgo sense of humour, but it
appears that he was unlovable on a personal level.
Neurotic Peter Sellers was only ever truly in love with
his mother, so much so that he plagued spiritual
mediums to keep him in touch with her after she was
dead. Like many Virgos, he liked sex well enough, but
was far too into himself and far too busy being famous,
being wrapped up with himself and with his
domineering mother to be much use as a husband or
father.

Vertex in Pisces

When the Vertex is in Pisces, the Anti-Vertex is in Virgo. Both are mutable signs, which suggest a certain amount of flexibility or adaptability. Pisces is a water sign while Virgo is an earth sign. Both Virgo and Pisces are signs that know how to sacrifice their own needs for others.

You may run a charity where you provide nurturing for those who need it, or you might become a religious leader or someone whose philosophy of life influences others beneficially. In such cases, you need a supportive partner who understands your goals. On the other hand, you may be a backroom person or a support system to a partner, children, parents and friends and colleagues. Whatever the scenario, your creativity and talent for expressing yourself will lead you to some kind of notability. You are a life-enhancing person who makes the world a better place.

You like to do things properly, to have a nice clean home and to be appreciated for the efforts you put in at work and at home but you also demand a lot of a partner in exchange. Your worst enemy is your tongue, because you can upset others with your frankness or your tendency to become angry, demanding and cutting when under stress. You can be surprisingly bossy but you can also suffer from bossy or unpleasant behaviour from others.

You probably fall in love quickly but you may not stay with any one person for very long. You tend to drift about as far as love relationships are concerned, looking for something better around the next bend. On the other hand, you may stay with your lover, with both of you leading somewhat independent lives. In the area where I live, there are many navy people or ex-navy people.

Their marriages are run on the basis of the man being away for almost a year at a time, then home for a few weeks and then away again. I remember a woman whose man was a lighthouse keeper. He would be away for a fortnight, home for a week and then away again. The Pisces Vertex would suit that kind of arrangement.

You may have a difficult relationship with one or both your parents and you may find your children something of a trial. The key to success for you and your partner seems to be flexibility – even to the point of turning a blind eye to the occasional infidelity. The relationship between you contains more than sex, so you need to feel that there is something deeper that keeps you together.

If you can share your interest in deeply spiritual matters with your partner, you will make a very successful team. If your restless feet take you travelling, you will either have to find a partner who is happy to travel with you, or one who is prepared to stay behind while you go exploring.

Politics may be your arena, in which case, your partner will need to hold the fort at home while you work long hours in Parliament.

Winston Churchill

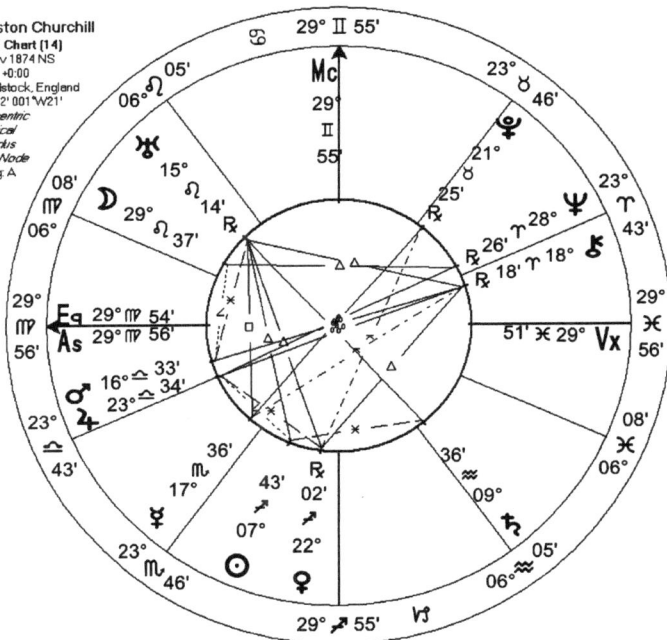

Winston Churchill
Natal Chart (14)
30 Nov 1874 NS
01:30 +0:00
Woodstock, England
51'N52' 001'W21'
Geocentric
Tropical
Placidus
True Node
Rating: A

Winnie was apparently known to have a flawed character. His wife certainly had a great deal to put up with, because Churchill was often moody, sarcastic, depressed or drunk. As a Sagittarian with a Pisces Vertex, he would not have had much idea about making or keeping money, so there was often a shortfall. His wife, Clementine, frequently complained about this, but she did nothing to bring in money herself. Winston often set to and wrote books to cover the debts. It is possible that his wife may have been unfaithful to him when she went on her mysterious sea cruises. Churchill was utterly faithful to her and loved her dearly. The marriage probably prospered, at least in part, because he was away from home so much of the time.

Julie Andrews

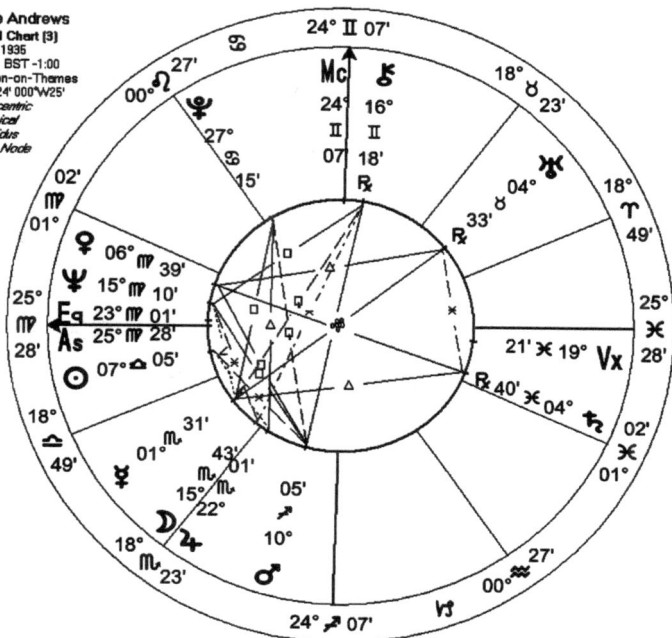

Julie Andrews
Natal Chart [3]
1 Oct 1935
06:00 BST -1:00
Welton-on-Thames
51°N24' 000°W25'
Geocentric
Tropical
Placidus
True Node

Beautiful, ethereal Julie Andrews brought so much pleasure and joy during her career as an entertainer. She has been happily married for many years to a supportive husband. Maybe the marriage also prospered because Julie was occupied with work so much of the time.

For someone with as lovely a voice as hers, it is tragic that she needed an operation on her throat recently, and she has been unable to sing. Hopefully, she may eventually recover, because, with the Vertex in the sixth house, her talent is important to her, even if she isn't using it at present.

7

The Houses

The Vertex through the Houses

The Vertex concerns relating, so it makes sense for it to live in the "relating" houses - the fifth, sixth, seventh and eighth. There are some rare cases when it is found in the fourth, and there may be others where it pops over into the ninth. Additionally, as in the case of Olga, who was born in Norilsk, in some cases, labelling in older versions of computer software will mistake the Vertex for the Anti-Vertex. For the most part though, the Vertex will be in the fifth, sixth, seventh or eighth houses.

The East Point

The East Point is the place where the Sun rises over the horizon on the celestial equator. All these points (Asc. Dsc, MC, IC, Vx, Anti-Vx, Ep and Wp) are nodes, i.e. places where one line crosses another.

The East Point can be in the same sign and house as the Anti-Vertex, with the West Point in the same sign and house as the Vertex. When this is so, it strengthens the influence of the Vertex and Anti-Vertex in the natal chart and makes these features more obvious when affected by progressions and transits.

Vertex in the Fourth House, Anti-Vertex in the Tenth House

Key ideas: Dynasty, status, position, money, prestige, politics, leadership and authority.

Someone born near or within the tropics can have a Vertex in this house. Indira Ghandi is one famous example. Her father was Jawaharlal (Pundit) Nehru, the father of modern India.

The fourth house refers to family matters, but not family life as we normally consider it, because here we see the "family" in the form of a dynasty. This might be a political dynasty, as in the case of the Nehru/Ghandi family or a dynasty of business, wealth or royalty. Women with this Vertex take on a power role, either within the family or in the wider world and these females can be impressive. Marriages are arranged for dynastic reasons rather than love. There is a sense of history here, so a child growing up with this Vertex would know what it means to be part of a dynasty.

If this is your Vertex house, you may be able to offset the lack of desire and attraction in your marriage by indulging in secret affairs, but, for the most part, your love will be reserved for your family and even for your pets. With luck, a dynastic marriage might turn out well, but while it may be that you can enjoy fame and fortune, yet you may never experience emotional happiness and emotional security.

The Anti-Vertex in the tenth house links with the ideas of status, position, prestige and dynasty. The father may have worn a literal or metaphorical crown, so this person is charged with keeping safe the hopes for the future of the family, tribe and nation.

Indira Ghandi

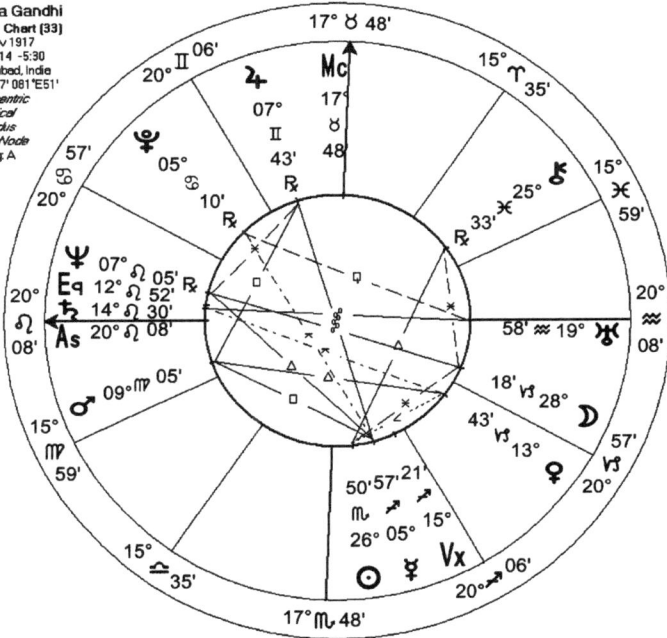

Indira Gandhi
Natal Chart (33)
19 Nov 1917
23:11:14 -5:30
Allahabad, India
25°N27' 081°E51'
Geocentric
Tropical
Placidus
True Node
Rating: A

A fourth house Vertex is unusual

Vertex in the Fifth House, Anti-Vertex in the Eleventh House

Key ideas: Romantic love, marriage for love, children and family life, glamour and excitement, entrepreneurs, wanting the best.

This is a common position for the Vertex and it represents the romantic love and the need to fulfil oneself by loving others. The affection wrought by this house can extend to partners, children, stepchildren and the wider family. You enjoy spending time with your loved ones, playing sports and games with them and talking with them. You need a soul mate, a commitment and a happy partnership. You have great intentions. The sexy nature of this Vertex can lead you into affairs, but what you are really seeking is love, rather than excitement.

You may choose a partner who is in a powerful position at work, who is nice looking or who works in a glamourous and exciting field. You need your lovers to be a little larger than life. You will put a great deal into all your relationships, including those with your family and friends, but you may not get as much back. On the other hand, you may be lucky - especially where your children are concerned.

This house links with entrepreneurial activities, so you and your partner may share an interest in business. You like to be among interesting and glamourous people, so you may work in a field that is exciting and prestigious. Alternatively, you might have a glamourous hobby that takes you into the world of spirituality or entertainment. Music, dancing, poetry, singing, ice-skating, sports and entertaining might fascinate you.

Natalie Cole

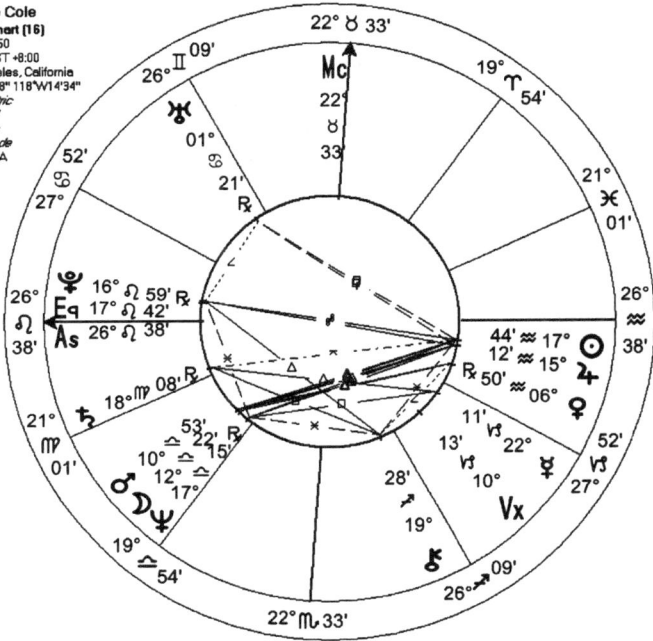

Modern astrologers tend to forget that this house was once also associated with a person's father, so it is possible your father is a great role model. You might choose to walk in his footsteps. I have found an excellent example in Natalie Cole for this house. Her father was the wonderful Nat King Cole, who became a household name around the world and a great success, despite being black at a time when there were few opportunities for African Americans to make a success of themselves. Natalie is a beautiful, glamourous woman and a wonderful singer.

When the Vertex is in the fifth house, the Anti-Vertex is in the eleventh house; the eleventh rules hopes and wishes, so perhaps this Anti-Vertex hopes for a little too much out of life and love. The eleventh house links to groups, clubs, societies and committees, so you may get a good deal of pleasure from being part of a group and from the friendship that this brings.

The issues here are those of dependence versus independence, also attachment versus detachment. You have a loving heart and a kind of spiritual generosity that lesser beings can latch onto. These people then lean on you or start to dictate to you; they think they can take your good nature for granted. You don't need to be propped up, you just need someone to share the load, that's all.

Vertex in the Sixth House, Anti-Vertex in the Twelfth House

Key ideas: Work connections, rescuing or being rescued, sickness, addictions or an interest in health.

You meet friends and lovers through work, and you might be one of those who falls for the boss and then loses the relationship and the job when the affair comes to an end. You may hire someone to help at work or in the house and then fall in love with him. If you don't work, you might find lovers through voluntary work, community work, interests and hobbies. Sometimes health is an issue, so your partner might be sick or disabled or he might need to be rescued from addictions or bad behaviour. On a happier note, you may both work in one of the many areas of health, or you may happily work together in some other field of endeavour.

Some of you prefer short-term affairs and long-term friendships to a conventional relationship. You value your privacy and you may find it hard to share your space with others. Some people with this placement are hard to live with, being boastful or apt to criticise.

As a rescuer, you provide a shoulder that others cry on and the rock that others lean on but this may never be reciprocated. Some of you move on from unloving parents to unloving partners and eventually decide that you are happiest alone. In some cases, relationships are not actually all that important because the career takes precedence. In this case, your emotions are expressed through your work and you get upset when things go wrong in that area of your life.

The Anti-Vertex is in the twelfth house, which explains the need for privacy, time spent alone and secret liaisons.

General Dwight D Eisenhower

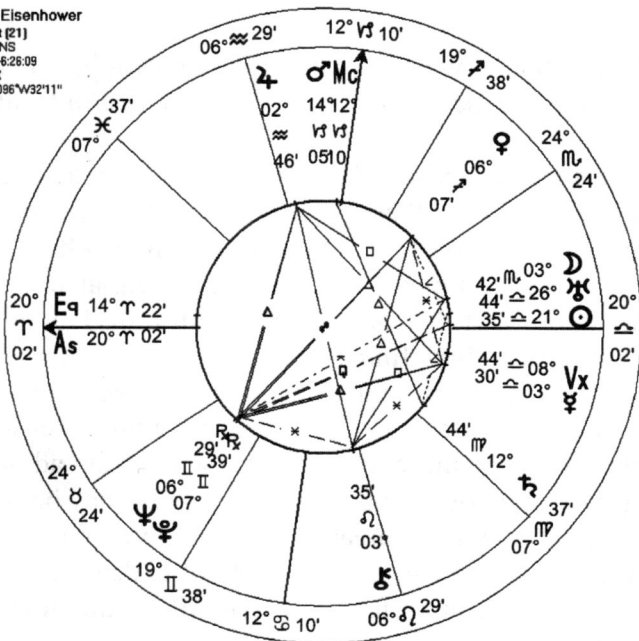

Dwight D Eisenhower
Natal Chart (21)
14 Oct 1890 NS
17:19 LMT +6:26:09
Denison, TX
33°N45'20" 096°W32'11"
Geocentric
Tropical
Placidus
True Node

General Eisenhower had a sixth house Vertex. He fell in love with a young woman called Kay Somersby who was his chauffeur while he was in England during the Second World War. Here, Eisenhower found someone he could confide in, relax with and be happy with, while coping with the mammoth task of being Supreme Allied Commander and the architect of the liberation of Europe. Once it was all over - surprise, surprise - he dropped Kay and went back to his wife.

Vertex in the Seventh House, Anti-Vertex in the First House

Key ideas: Luck in love, needing partners, learning to stand up for yourself, dependence, music and the arts.

The Vertex can be at its most comfortable in this house, so falling in love, marrying, having children and making friends come naturally to you, and you may be luckier in relationships than most. You are probably nice looking and you may be flirtatious, charming, kind, well balanced and helpful. Your issues surround dependence and independence. You may attract a partner who leans on you and wears you out or you may lean too heavily on others. Having said this, you learn to like your own company later in life, so after years of giving your emotional support and energies to others, you may end up alone. This would be more by choice than by circumstance, but you can only stand this if you have many friends and interesting occupations or hobbies.

You might enjoy music, art, acting, performing or being on show in some way or you may be attracted to those who are in these fields. If you love a performer, you may subjugate your needs in order to promote his talents. Alternatively, you may spend a lot of time sitting around and waiting for him to finish work or waiting for him to come home.

The Anti-Vertex in the first house means that you may be eager to hand over power to others or you may fight to keep the power in your own hands. On a nicer note, you want to do something worthwhile during your life, possibly teaching or working for the benefit of the public. Oddly enough, the career can be as important to this person as it is for someone with a sixth house Vertex.

Larry Hagman

Larry Hagman
Natal Chart (24)
21 Sep 1931
16:20 +6:00
Fort Worth, Texas
32°N43'31" 097°W19'14"
Geocentric
Tropical
Placidus
True Node
Rating: AA

My celebrity for this Vertex is the loveable bad man, Larry Hagman. He played the fascinating, sexy, manipulative, crafty JR in the Dallas series. In real life, he is charming, amusing, clever and clearly a very nice person, although when he was younger he spent years drowning himself in drink. I have never heard any sexual scandal about him and he even appears to love his larger-than-life mother, Ethel Merman, who popularised the show-biz anthem, *"There's No Business Like Show Business"*.

Vertex in the Eighth House, Anti-Vertex in the Second House

Key ideas: Sex, a sexy image, passion for something, interest in psychic and spiritual subjects, struggles over money, power, control, dirty tricks, bereavement or more than the usual amount of contact with death.

There is a level of intensity here that can take a variety of routes. Relationships may bring heights of love, passion, affection and sexual fulfilment and a sense of deep commitment. You seek out deep thinkers and those who you hope also have deep feelings, but far too often, you choose badly. You learn the meaning of negative emotions such as jealousy and obsession and you may even go through periods where you feel humiliation, bewilderment, loss, rage and bereavement. However unpleasant these emotions are, it is not a good idea for you to bottle them up. There are some very good sides to this Vertex house, so you can expect to feel great highs as well as lows during your life. Somewhere along the line, you will experience some really great relationships, perhaps with friends, children or perhaps in a subsequent marriage.

You have an attractive, charismatic image that brings people into your orbit. Your sexy image might take you into acting, the media or into making the world a better place. You may be interested in politics and in those who are successful in business, entertainment or any variety of positions of power.

You attract those who seek to control and dominate you, perhaps telling you what to eat and drink and trying to prevent you from smoking. This may lead you to take a destructive, druggy path just to spite them. You may become promiscuous, either because it upsets your parents and family or in an attempt to find love. You may wish to control others, with the obvious result that they leave you. There may even be an element of masochism working here. For example, this is the kind of person who may fall in love with a married man who keeps her on a string while avoiding commitment.

It may be hard for you to find love, friendship and sex in the same place. You will have extremely close or extremely bad relationships with parents, siblings, friends and colleagues. You may idolise parents, relatives and your lover and then feel very let down when they show themselves to have feet of clay.

The Anti-Vertex is in the second house, which rules personal values and priorities and personal finances. It might be hard for you to hang on to money while other people are in your life. Alternatively, ownership of money and goods might be triggers for abuse, domination and control of one partner by another.

Lucille Ball

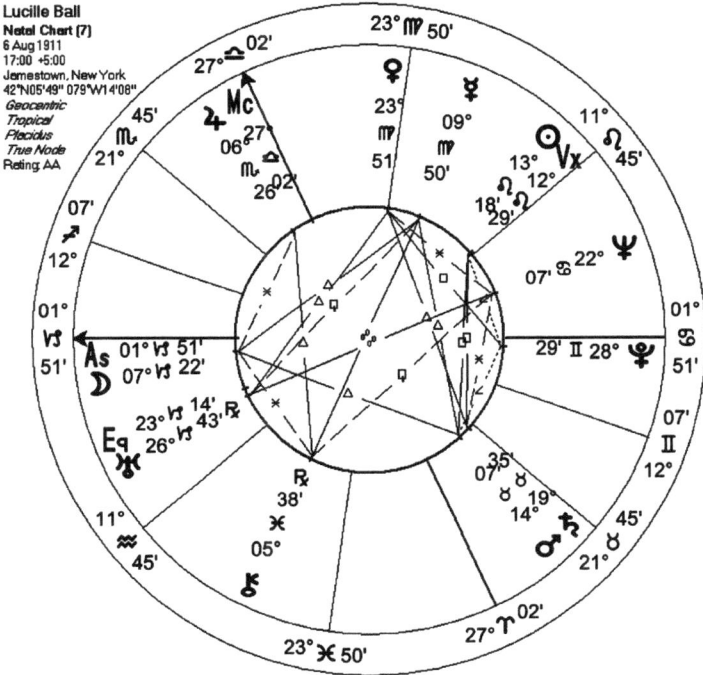

Lucille Ball
Natal Chart (7)
6 Aug 1911
17:00 +5:00
Jamestown, New York
42°N05'49" 079°W14'08"
Geocentric
Tropical
Placidus
True Node
Rating AA

My eighth house example, Lucille Ball, was the business powerhouse behind the amazingly successfully Lucy and Desi TV show and the film, "*The Long Trailer*", in which she acted with her husband, Desi Arnez. Desi allowed Lucille to make him rich; then he carried on with other women until Lucy got fed up with it and dumped him. She then went on to make a successful career for herself, both in front of and behind the scenes.

Vertex in the Ninth House, Anti-Vertex in the Third House

Key ideas: Desire for freedom, expanding horizons, an interest in travel, foreigners, spiritual matters, studying and teaching.

This is such an unusual placement that despite searching my database, I can't find one example, so it is a matter of guessing what this might mean. Religion and belief might take the place of relationships, so this person might be a high priest who is loved and looked up to by his flock. He may have friends among his courtiers - and backstabbers there too, no doubt - but personal relationships may be sacrificed in favour of his position. There is a certain detachment with regard to relationships. The chances are that this person doesn't live in the place where she grew up, and that she detaches from her family at an early stage in life. She may travel a great deal, and she may have many friends, but few who are really close.

This person will be a writer, teacher, counsellor and religious, philosophical, medical or academic leader. The Anti-Vertex is in the third house, and this also talks of intellectual interests, education and a quick and active mind. She is attracted to friends who are as quick and clever as she is. She may be unusual or eccentric.

8

Aspects

I don't think that natal aspects to the Vertex or Anti-Vertex are desperately important, apart, perhaps, from a tight conjunction, when the planet in question will have some bearing on the way that unmanageable things happen to a person. In that context, the most interesting of these would be a conjunction between Uranus and the Vertex.

The following list can be used as a starting point for your own research. I suggest that these relationship connections can be helpful and fulfilling, or disappointing and painful.

PLANET IN CONJUNCTION	PEOPLE WHO HELP OR HURT
The Sun	The person, her children or her father.
The Moon	The person, her mother, her family.
Mercury	Her siblings and neighbours.
Venus	Her lover, partner, female friends or relatives.
Mars	Her lover, partner, male friends or relatives.
Jupiter	Her teacher, guru, someone she believes in.
Saturn	Her father, other older relative or friend, boss, authority figure, person at work.
Uranus	Her friend, group, society, local government, teaching and other organisations.
Neptune	Someone who is kind and supportive or who betrays her, spiritual people, musicians, artists.
Pluto	Any relative or other person for whom she has strong feelings, her psychiatrist.
Chiron	Her teacher, helper, medical and other healer.

9

Synastry

Those astrologers who have studied the Vertex have discovered that it frequently links with an axis or planet on another person's chart. Thus, the Vertex will be close to the other person's Asc, Dsc, MC, IC, or perhaps their Vertex, Anti-Vertex or East or West Point, or to a planet. You might say that there are plenty of features to choose from, but when you actually come to look at charts, the chances of finding something within a few degrees of a conjunction are not that high.

The following statement will probably outrage many astrologers, but I suggest that, when using synastry, you give the Vertex at least a ten-degree orb for important relationships and love affairs. When considering friends, business acquaintances and so on, use the Vedic "whole sign" principle. Naturally, the closer the conjunction the more effective it is likely to be.

The person who discovered the Vertex in astrology, Edward L Johndro, believed it to relate to fateful encounters and circumstances. He believed that it connected with uncontrollable events that involve other people. I know one person who has had many Vertex connections with lovers and with others in her life, and she has outlived them all. This means that she has been bereaved many times.

We think of synastry as a love connection, but it can relate to any kind of connection. For example, one of my friends had the mother-in-law from hell the first time round, and a lovely mother-in-law the second time around. The first had an ascendant that linked with my friend's Vertex and the other had a Sun that made the connection. It seems that the thrusting nature of the ascendant sought to crush my friend, while the warm, loving and encouraging Sun sought to love and support her.

A Set of Famously Painful Vertices

Early in 1981, I was doing some housework with the radio playing in the background and I heard the presenter announce that the next item would be an interview with Prince Charles, ahead of his forthcoming wedding to Princess Diana. I turned the radio up and was astonished to hear what the Prince had to say. A strange mixture of apathy and desperation seemed to seep out of the wireless. Prince Charles talked about himself in the third person in that weird way that the Royals do, but it felt to me that he wanted to be someone else… anyone else rather than himself.

He answered the interviewer with words something like, "The heir to the throne must marry someone who has no past…". He also said something to the effect that the Palace had trawled Europe's royal princesses to see if there was anybody he could stand the thought of living with, but that he had eventually chosen Lady Diana himself. It struck me that the great romance the newspapers were making of the situation was absolute rubbish and that he was simply being married off to a suitable virgin. I had a horrid vision that the poor girl might even have had to submit to having her private parts inspected to prove her virginity, as Queen Elizabeth the First once had to do.

Later it transpired that it was not Charles who chose Diana, but Camilla Parker-Bowles. Charles and Camilla must have thought Diana really stupid and controllable. In effect, Prince Charles was already very happily married - to Camilla. His poor, naïve, teenage fiancée was led to believe that the marriage she had bought into was real. This dream was shaken when the strain of preparing for the wedding sent poor Charlie running into the arms of Camilla. Not everything has a downside though, and Diana suddenly found herself fabulously wealthy. She didn't have to go out to work, clean a house, wonder what her hubby wanted for his supper or change nappies. Indeed, she could spend her time dancing, swimming, having her hair done, designing her next wardrobe and picking out suitable jewels to go with it.

In the long term, the whole business took its toll on everyone concerned, and to some extent, even Charles suffered a bereavement of sorts when Diana was killed.

How ironic that it was only when Charles and Camilla were standing on the cusp of old age that they could marry and be content together. What a strange pair they make, with their stuffy Edwardian clothes and weird walking sticks…

Prince Charles's Vertex is 22 deg. Sagittarius, his ascendant is 5 deg. Leo and his East Point is 11 deg. Cancer.

Princess Diana's ascendant is 18 deg. Sagittarius and her Vertex is 4 deg. Leo and her Sun is 9 deg. Cancer.

Duchess Camilla's Vertex is 20 deg. Sagittarius, her ascendant is 3 deg. Leo, her Moon is 9 deg. Cancer and her East Point is 8 deg. Cancer.

Charles, Prince of Wales

Prince of Wales, Charles
Natal Chart [13]
14 Nov 1948
21:14 +0:00
Buckingham Palace
51°N30' 000°W08'
Geocentric
Tropical
Placidus
True Node
Rating: A

There are many more connections to be found in these three charts, and probably enough for a book in itself about their progressions, solar arc, transits and goodness knows what, but keeping to the remit of this book, it is interesting to note the following:

Prince Charles's Vertex in the fifth house suggests that he is more loving and lovable than we might think. Perhaps he was as much a victim as both his wives were.

Duchess Camilla

Camilla Parker-Bowles
Natal Chart (29)
17 Jul 1947
07:00 -2:00
London, England
51°N30' 000°W10'
Geocentric
Tropical
Placidus
True Node
Rating: B

Duchess Camilla's Vertex is in her fifth house of love affairs and children. Her West Point is in the sixth house of duty. She loved deeply, but she did her duty as well, waiting so long before she could live openly with the man she loves.

Princess Diana

Princess of Wales, Diana
Natal Chart [20]
1 Jul 1961
19:45 -1:00
Sandringham, England
52°N50' 000°E30'
Geocentric
Tropical
Placidus
True Node
Rating: A

Princess Diana's Vertex is in the resentful eighth house of shared resources, wealth, sex, stupid affairs, "transaction" marriages – and untimely death. Her West Point searches for everlasting love in the seventh house. Her lovers used her, but she found true love among the millions of people in the world who adored her.

On the day of Diana's death, her progressed Vertex hit her Mars in Virgo, and that in itself is close to the midpoint of her natal north node and Pluto. A nasty day indeed.

Vx

10

Forecasting

You can use all your favourite forecasting techniques when dealing with the Vertex, including progressions (secondary directions), transits, solar arc, solar and lunar returns and so on. You can use horary, asteroids, fixed stars, Arabic parts, midpoints, lunar mansions, Decans, Dwaads (see my book "The Hidden Zodiac" for full treatment of Decans and Dwaads) and anything else that takes your fancy. I have found that the best methods for most purposes are good old progressions and transits.

Remember that the Vertex is active when dramatic situations occur that are outside the normal run of life.

Transits *by* the Vertex

The Vertex itself can make transits, but it is linked to the ascendant, so it travels around the whole chart during the course of a day. This means that the transiting Vertex won't have much effect. It might be worth plotting its progress round your own chart for a day or two, to see whether it brings moments of pleasure or irritation as it passes by sensitive points.

I have just looked at my own chart and noticed that within the next hour, transiting Vertex will conjunct transiting Pluto while it trines my natal Sun and then opposes my natal Saturn. This means that the Anti-Vertex will conjunct my natal Saturn, sextile my Sun and oppose transiting Pluto. I will keep you posted later in this chapter and let you know if anything (or nothing) happens. Talk about astrology in action!

Transits *to* the progressed Vertex

The importance of transits to the progressed Vertex or Anti-Vertex will depend upon the planet in question. For instance, a lunar conjunction will pass within a couple of hours and it happens every month, while a Pluto conjunction will take several months to pass and it can only happen once in a lifetime.

For the sake of this book, I have just checked my own Pluto transit over the Vertex. This was exact in December 1987, then by retrograde motion in April 1988 and again by direct motion in October 1988. The years 1987 to 1988 were among the most successful and happy times in my life. Some of my most successful books were published during that period, and for the first time ever, I was earning decent money. There was one weird event in December 1987. An old flame contacted me and invited me out to a meal "for old time's sake". Despite some trepidation, I agreed to see him, but the evening turned into a disaster, as the man got very drunk and unpleasant, and I never saw him again after that.

Nothing sticks in my mind about April 1988, but October 1988 turned out to be life changing. In the autumn of 1988, my publisher asked me if I would like to go to South Africa for a book tour and to give lectures at a mind, body and spirit festival. This meant leaving the family and going halfway around the world for almost a month. It meant being on show and being examined by people that I didn't know from Adam, so I judged that it could be quite an ordeal. Despite feeling nervous about the idea, I decided that it could only do me good.

As it happens, I had the time of my life and I was appreciated and accepted by everyone in South Africa. I made firm friends with Vivien, the sales manager from the South African distributor, Alternative Books, and I became part of her family for a while. Vivien now lives in the UK and we still laugh about things that happened on that visit. It was my friendship with Vivien that took me back to South Africa eight years later, and that was how I met my husband, Jan.

Now I must report back about the Vertex transit that I mentioned earlier in this chapter. Nothing significant happened, apart from the fact that transiting Vertex crossed transiting Pluto while I was mentally reviewing a time in my life when transiting Pluto crossed my natal Vertex.

Transits: A Small Cookbook

Obviously the transits that matter are those of the slower moving planets, but astrologers who study the Vertex have noted that transits can pass by with nothing of any particular importance happening. I have noticed this pattern with quite large planets at times (especially Jupiter), so the Vertex is not alone in this respect. I think it all depends upon a variety of factors, such as whether the stars and one's karma happen to line up at that time. I think the only transits that are worth considering are the major ones. The following is what you might expect to happen - among many other scenarios…

TRANSITS TO THE VERTEX	
The Sun	A pleasant/unpleasant encounter.
The Moon	A pleasant/unpleasant encounter with a family member.
Mercury	Someone gives you good news or a good idea. Someone upsets you or you upset someone.
Venus	A pleasant/unpleasant encounter with a woman. A romantic or hurtful gesture that others make to you or that you do to others.
Mars	Someone rushes to your aid. A driver cuts you up and gives you the finger.
Jupiter	New people enter your life for some purpose – good or ill.
Saturn	Depressing people enter your life. Either they are a pain in the butt or they help you to get some structure in your life.
Uranus	A sudden revelation, an epiphany, being rescued or dumped in the proverbial compost heap.
Neptune	A love affair that is memorable, but with someone who only tells you what he wants you to know. If this ends badly, it will be extremely painful.
Pluto	Life changing times or events.
Chiron	An illness or accident forces you to reassess your life and change it for the better.

A Personal Story

I was a professional performer and entertainer from the age of twelve onwards, but eventually I got fed up with the travelling life. When I was 18, I settled in one place and took a job in a shop that sold musical instruments, music and records. I had always loved music but this job allowed me to really get into classical music, which I have loved ever since. The shop was part of a large chain and Tony Fenton was one of the supervisors. He asked me out and soon we became engaged, and we had a great time together. I saw all kinds of wonderful things in Tony that didn't really exist, so after we were married, I came up against the fact that the Tony I thought I was marrying was not the reality. Neptune was conjunct my Vertex at the time, and that accounts for my rose-coloured spectacles and being surrounded by music – and being happy.

Finally, we must remember that the Vertex can talk about work output. You may remember that I mentioned that 1987 and 1988 were particularly productive times for me. That is another typical Vertex factor - that of success in business or at work.

Some Techie Data

The transiting Vertex moves at an erratic pace. It can speed up, slow down or even turn around and start to move in the opposite direction, all within a few hours. This isn't the same as a planet turning to retrograde motion.

In most charts, the Vertex normally moves up the right hand side of the chart from the fifth house to the sixth, seventh and eighth house, then it turns tail and travels back down from the eighth, through the seventh, sixth and fifth before turning forward again. In births on the equator and extreme locations near the north and south poles, it will move until it meets either the MC or the IC, when it will suddenly switch from one end of the chart to the other! If your software allows you to see the horoscope in motion by means of an "astrological clock" feature, run a few charts and see what happens, using different time intervals, from a few minutes, through hours, to days. Try this with births in various parts of the world to see how the different Vertices work.

Remember, the transiting Vertex moves quickly, so it doesn't have much impact on a person's life. Its action is similar to that of the transiting ascendant or MC, and much less than the Moon. It may cause an hour or so of amusement or irritation, but that's about all.

Eclipses and the Moon

Eclipses are far more important than many astrologers realise. An eclipse on any planet, angle or sensitive point in a chart will always make itself felt, while an eclipse on the Vertex or Anti-Vertex could bring a relationship, job or some other long-term situation to an end. Bear in mind that a good relationship won't fall apart due to an eclipse, but a difficult one might very well be pushed off its perch.

If an occultation happened on or near the Vertex or Anti-Vertex, this could be really memorable, and not for the nicest reasons! An occultation occurs when the Moon passes in front of a planet. The way to see it is to look at the lunar aspects in Raphael's Ephemeris. When you see a symbol that looks like a conjunction but with the middle filled in, this is an occultation. If it is in the Sun's column, it is a solar eclipse. An eclipse and an occultation are the same thing - eclipses simply refer specifically to the Sun and the Moon, while an occultation refers to any planet that is temporarily hidden by another celestial body.

A new Moon brings a time of small beginnings, while one on the Vertex will bring some minor fresh start or fresh outlook in connection with others. The full Moon will bring tension and trouble, and it could be quite nasty if it happened exactly on the Vertex or Anti-Vertex.

Time and again, I read something like this in some woman's magazine: "The full Moon on the 20th will bring you a stunning opportunity for romance or for finding a windfall." What twaddle! The full Moon brings tension and aggravation to whatever house it falls in.

Progressions (Secondary Directions)

This is where the progressed Moon and faster moving planets come into play. If a planet progresses to an aspect with the Vertex, it will mark a memorable period. Naturally, the Vertex also progresses, and the progressed Vertex can really make itself felt. Watch for those times when the Vertex changes sign or house, as this can be interesting - although not always in the way that one would expect. Here are some of the effects that I have observed.

PROGRESSED ASPECTS TO THE VERTEX	
Progressed Mars trine Vertex	A passionate affair
Progressed Venus square Vertex	A star-crossed affair that should never have happened
Progressed Venus square Vertex	A marriage coming to a long, slow, painful end
South node conjunct Vertex	A separation
Jupiter conjunct Vertex	Meeting and later marrying someone nice
Jupiter square Vertex	A lover leaves and goes overseas
Vertex changing sign	A child is born
Vertex progressed to the descendant	The start of an excellent business partnership
Pluto square Vertex	A death in the family
Eclipse on the Vertex	In one case, a damaging argument and in another, separation followed by divorce

Remember that the Vertex can talk about output, so if the person concerned is interested in her work, business, a creative hobby or a fund-raising event, look at the Vertex for times when she will make a success of things. Also, look for the following:

> Times of joy
> Childbirth
> Falling in love
> Heartbreak
> Bereavement
> Finding a soul mate
> A hateful person entering one's life
> An important house move or change of location
> An important journey
> An accident
> An important meeting
> Any major event that is out of the normal run of things

Solar Returns

The solar return marks the flavour of the year in question. An aspect to the Vertex in the return chart will show how major changes will take shape during the coming year. Relationship changes are the most obvious thing to look for, but there might be a change in outlook, the discovery or loss of a teacher or guru, a change of job or a massively good or poor result after a lot of hard work.

Lunar Returns

If you fancy looking into these, you can apply the same idea as for solar returns – in that an aspect to the Vertex will show what will happen overall during the month in question.

Other Methods

If I were to choose a time to start something new and create an election chart, I would certainly not wish to pick a time when something nasty was happening to the Vertex or Anti-Vertex.

11

Conclusion

I would never have looked into the subject of the Vertex if I hadn't gone to South Africa in 1988. One of the people who had booked me for a lecture suggested that I give a talk on the Vertex. The woman who arranged this particular talk had come across it in a booklet written by Maritha Pottenger and Zipporah Dobbins, and for some reason, she expected me to be conversant with it. I'd never heard of the Vertex, so I immediately began looking into it. When I discovered what it was and how it worked, I became totally fascinated with it.

In this small book, I have written all I know about the Vertex, so now I leave it to you to look at it for yourself and to check your own findings. Look back at those times when you were in love or in trouble, and see what was happening.

Index

Zambezi Publishing Ltd

We hope you have enjoyed reading this book. The Zambezi range of books includes titles by top level, internationally acknowledged authors on fresh, thought-provoking viewpoints in your favourite subjects. A common thread with all our books is the easy accessibility of content; we have no sleep-inducing tomes, just down-to-earth, easily digested, credible books.

Please visit our website at *www.zampub.com* to browse our full range of astrology and other MB&S titles, and to see what might spark your interest next...

All our books are available from good bookshops throughout the UK; many are available in the USA, sometimes under different titles and ISBNs used by our USA co-publisher, Sterling Publishing Co, Inc.

~~~~~

Please note:-

Nowadays, no bookshop can hope to carry in stock more than a fraction of the books published each year (over 133,000 new titles were released in the UK in 2004!). However, most UK bookshops can order and supply our titles swiftly, in no more than a few days. If they say not, that's incorrect.

You can also find all our books on amazon.co.uk, and many on amazon.com. Our website also carries the whole range, including our latest innovation, the first six titles in the "Simply..." colour-illustrated series.

Reading the Runes

BAPS (The British Astrological and Psychic Society) and Zambezi Publishing have developed certain projects as standard texts for BAPS certified courses.
If you are interested in studying for a recognized qualification in Rune Reading, or a number of other interesting MB&S subjects, then you may wish to contact BAPS for further details. Their address is shown below.

~~~~~

These are some the courses available from BAPS, some of which are accompanied by Zambezi Publishing text books:
Psychic perception ~ Astrology ~ Classical Astrology ~ Karmic Astrology ~ Tarot ~ Palmistry ~ Chinese Oracles & Feng Shui ~ Crystal Divination ~ Dream Interpretation ~ Graphology ~ Numerology ~ Practical Witchcraft & Magic ~ Introduction to Alternative Health.
(The Course list may change from time to time - contact BAPS for the latest details).

~~~~~

Please address enquiries to:
Dept Z
British Astrological and Psychic Society
P.O. Box 5344
MILTON KEYNES MK6 2WG

~~~~~

Tel: +44 (0)906 470 0827
web: www.baps.ws   email: info@baps.ws